Mixing for God

A Volunteer's Guide to Church Sound

Barry R. Hill

Mixing for God: A volunteer's guide to church sound

Cover design by Scott Cole, Churchill Creative

Printed in the United States of America

ISBN: **978-1535189415**

The audio examples are available at www.barryrhill.com/mixingforgod.

CONTENTS

Mixing for God

Good for you. You felt God's calling to get involved in your church's sound ministry. Maybe you've run sound for your band. Maybe you struggle to play the FM radio. You might be one of the rare ones who have an engineering background in audio. Either way, you need to get up to speed on how to run all of this expensive equipment to help make sure your church's services run smoothly. Even though there's a lot involved, you don't need an engineering degree to do a great job. You don't need to know how to design systems or solder patchbays. You *do* need to understand how things basically work, how to handle situations that come up, and you need to develop a decent ear so you can put together a musical balance of your musicians and vocalists. Poorly run sound is at best a distraction from the worship environment and can destroy that experience altogether. Your job is to make all of the technical work transparent so that nobody realizes it's there at all. When a mic is not turned on in time or feedback starts howling across the room, people stop and notice. When the mix is right, people stand and worship.

I've tried to present chunks of information to help you learn how to get this done without overloading on the technical background. If you're new to all of this, you'll find a warm and fuzzy comfort zone to guide you along. For those of you who've spent years running sound, you'll find new ways of understanding the complex world of audio, acoustics, and music that should improve your mixes and operational understanding. So, you don't necessarily have to read cover to cover—just jump wherever you need help.

We start in *Level 1* with the big picture, giving you an overview of the entire mix system, briefly describing the equipment and primary operations involved in running sound so you'll know how everything we'll be discussing works together. Then in *Level 2* we start setting things up for services, getting you up and running on a Sunday morning. You'll learn how to set up mics and get sound through the console.

Once you get everything going we can work on improving your mix, solving typical problems, and providing some technical background. *Level 3* fine-tunes all of this with more ideas on how to select and place microphones and get a really good mix, understanding the various parameters involved in blending all of your sounds together. If recording your services is of interest to your church, I'll briefly describe some options and how this works in conjunction with running your live mix. Since your week just wouldn't be complete without its share of frustrating problems and weirdness, you'll find comfort in browsing through a long list of solutions to most of these. When you want more technical explanations on things such as EQ and acoustics, drink lots of coffee and slog through *Level 4*.

Finally, you've got to train your ears so you know what to listen for, so follow the audio examples sprinkled throughout the book. You've got to know when the guitar is too loud or the alto vocal is lost in the mix. Play the audio examples on a good quality system so you can hear bass, treble, and everything in between. Listen to them over and over while following what the text says about each example. Over time you'll be amazed at what you can hear and, more importantly, really impress your friends at next year's Christmas Eve party.

Level 1
The Big Picture

Level 2	
Setting Up	
On stage	On the console
Power up Setting up mics	Main mix Monitor mix

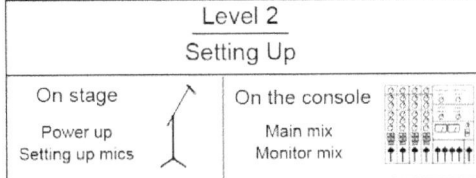

Level 3
Refining your sounds

On stage: mics & acoustics	On the console: mixing & processing
Which mic to use? Which way do I point it? How close do I put it? Acoustics issues for miking How many mics should I use? Wireless or wired? Mic selection & placement	What's a mix? How do I do all of this? Musical balance Tone Reverberation Uneven signal levels Noise from open mics Distractions

Level 4
Digging deeper

Mixing consoles	Cables	Console connections	Microphone design	Microphone technique	EQ	Room acoustics	Audio fundamentals

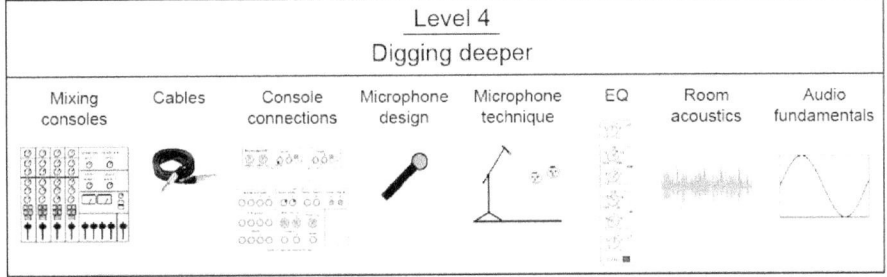

The consultant lends a hand

3rd Street Covenant Fellowship is located in the heart of North Carolina. It's not a big church by anybody's standards, especially when compared to today's mega-blockbuster operations. They top out around 280 for a typical Sunday, maybe 325 when families show up for Christmas, and even higher when there's free ice cream. It's a well-organized church, though, with active ministries doing great things in the local community. What they don't have is an experienced sound engineer to run their services. Jonathan is their main go-to sound guy with a track record, which means he used to play in a rock band in college before he went on to medical school. Between emergency calls to the hospital he oversees a technical team of two—himself and Madison, a high school sophomore who would really like to play in a rock band in college, but hasn't yet figured out which instrument he'd like to learn.

Jonathan somehow found out that you were pretty good at this sound stuff and was available (and free) to come help them out. So, you've agreed to spend a Sunday morning and see what was going on as they set up and run through a service. This is where our story begins—fairly iffy at first, quickly beginning to wobble off the rails, then going horribly wrong...

Sunday, May 29...Setup

A slightly rumpled Jonathan eagerly greets you at the front entrance before leading you into the sanctuary. "I've been up all night delivering babies, but I'm really glad you're here. I think we need help."

"Glad to be here. Are the TV screens in the lobby for live streaming the service, or just signage?"

"Both, so when the service starts we run a video feed around the building so everyone can watch. There's a camera up on the back wall."

The main room is fairly rectangular in shape, wider than it is long, so people aren't sitting too far away from the stage area. There's a slight steam-punk feel and look to the room. The bronze color carpet covers the entire floor, and the walls sport a subdued, warm white paint job from end to end. The theater-style seating is plush and looks really comfortable, way better than the hardwood pews you grew up with. About 18' or 20' up, the ceiling is flat and covered with those standard acoustic tiles you see in office buildings. Front and center hangs the main (actually the only) speaker array for the room, with five cabinets facing an arc around the room. The band is up on the left side of the stage, choir risers are located to the right, and two large overhead screens are mounted on the front walls. The pulpit is an interesting retro-industrial metal design. Not sure what catalog that came from, but probably not where they buy the communion cups.

Up front, Madison is busy plugging mic cables into the stage panel. The rest of the worship team has not yet arrived, waiting, of course, for the tech crew to get things together first. Jonathan walks over to the sound booth and flips a switch to turn things on...

BOOOMMM...—a shrug of the shoulders—"it always does that"—and he moves on to something else.

"Hey Jonathan" Madison calls from the stage, "Which number is the acoustic again?".

"Seven."

"OK, thanks."

They're running a small, but decent digital console, and you scan it looking to see how they've got things set up. 24 channels, all assigned to the mix bus. Eight aux sends, plenty to work with, but only a couple of them seem to be used. An iPod sits over to the right.

"Which aux is running your stage monitors?"

"One is for the band, two for vocals, although the band mainly uses their new personal mixers. Man, they're really nice."

Another CRAAACCKK. Madison just plugged another mic in. He's almost done, looks like.

The band's beginning to drift in and get set up. Let's see, drummer, keyboard, bass, acoustic guitar, electric guitar dude dragging a humongous amp across the new carpet, and even a small horn section—trumpet, sax, t-bone. Very cool. They've obviously been doing this awhile, so they plug right in and start warming up. Direct box on the keys, acoustic, and bass. Madison has strung a 57 hanging down over the guitar amp, which is on the floor facing out toward the room. Horns stand over to the side looking toward where the singers will be, with one  of those small choir mics on a stand close in front of them, about 5' high. The drumset is surrounded with your standard plastic drum shields, the 4' high ones. Looking through the panels you can see a kick mic facing the front of the drum, snare mic pointing at the edge, and tom mics mounted with those special drum clamps.

While they get in place you walk back and glance around the control booth. Nice shelves on the back wall—stacks of equipment manuals, spare batteries, and old bulletins, with the wireless receivers up top. The entire booth is elevated a couple of feet, which is really nice to help you see a bit better. Lots of wires running everywhere, though…here come the singers, coffee in hand. Looks like Jonathan is about ready to fire things up.

Sunday, May 29…Sound check

Here we go. Bass guy is cranked and ready. Subs are great here—good and loud. Boomy, though, but then the room is still pretty empty. Acoustic guitar is having a problem—no sound. Jonathan checks channel seven to make sure it's turned on. Fader up, but nothing's flashing. Hmm. "Madison, you plugged acoustic into seven, right?"

"Yeah, I'll go check." Looks like everything's plugged in.

"Let's move on for now. Singers! One at a time…wait a minute, Bob, which mic are you holding?"

"Uh…12."

"That was Shelley's mic last week. Hang on while I change a few things. Say a few words." He quickly pulls the fader down and dials the low EQ down a bit since Bob's voice is louder and has more bass. "Next". Everyone's now on and working.

Not wanting to get in the way, you finally lean over and ask "You hear that?"

"What? Oh that hum? Yeah, it's always there. You won't hear it for long. Ok, guys, go for it."

"Two—three—four"…and the Grateful Dead's Wall of Sound has been resurrected. Boy, it's loud. One look at Jonathan shows he's frantically yanking faders down by the handful. Now, move a couple here and there a bit more carefully as he tries to pull it all together. Ok...you can hear a band slowly emerging. Except for the acoustic guitar, of course. Still out. Lots and lots of electric guitar over everything else. Bass booming around, and the horns so edgy and shrill they're peeling your eyebrows off. And that's just the band.

First verse. No vocals. "I thought I had them a moment ago." Up go the faders, and *then* they begin singing now that they've found their place in the song. Down go the faders. Ok, that's good.

"Got it. Thanks."

Really? The worship team keeps rehearsing their songs for the service, but Jonathan and Madison turn toward more urgent matters, like typing in all the overhead announcements, scripture, and other stuff for the video display. And more coffee. After all, the service starts in about 20 minutes.

Sunday, May 29...The service begins

The iPod is there to play pre-service music while folks arrive. The tech team is still busy typing, adjusting fonts, and greeting friends as they enter the sanctuary. Meanwhile the worship team is also waving to their friends...no, wait...they're flagging someone down. Probably Jonathan, who needs to turn their mics on. Finally the drummer kicks off the first song, but all we hear are muted drums, piano, and a really loud electric guitar. After several bars the mics come alive, and immediately all conversation in the room gets louder as people strain to make themselves heard over the din.

Once folks realize the service has started, they stand and sing enthusiastically. A couple songs are fairly new, and they can't quite hear the vocal melody to follow along. Half the congregation tries to follow the horns, which are still slicing through everyone's sinuses. It goes okay, mostly.

"What's that?" A low rumble, barely noticeable for the past eight minutes, has finally caught their attention.

"Feedback," you offer helpfully.

"Yikes. Got it," pulling down the high frequency EQs on the vocal mics. Still there. Attenuate more. Still there. Pull faders down. Gone now. So is all intelligibility and volume from the vocals.

The electric guitar fader, of course, is now all the way down. If he had a utility knife he'd try to cut a longer notch in the console to go further. And it's still too loud. And still no acoustic guitar. Bass guitar, however, sounds pretty decent. It's still muddy in the room, but at least it's strong and the player is really good. The piano is fairly easy to hear since it's a real one, not electronic, so it carries well.

While all this is going on, you decide to scoot out the back and wander around the building. The TVs are indeed pretty cool, and people can watch everything going on in the sanctuary while taking care of whatever business they're tending to. They're all muted, though.

"Hmmm..." You reach over and turn up the volume. Oh my. Better leave it down. One, maybe two vocals way out front with a piano far in the background. The rest of the band sounds more like, well, hash or something. Walking by you notice a speaker in the ceiling for the nursery, but again no sound. Being the expert-on-the-scene you helpfully turn it up...nothing. At all. "Oh, it's been broken forever." They don't seem to mind, having given up all hope long ago.

Back to the sanctuary. The band is good and the singers are spot-on. At least what you can hear of them. Poor Jonathan is struggling to stay alert, having had no sleep the night before. He was right, though. That hum is nowhere to be heard. Madison is running the overhead slides. The last song ends, a few thuds as the vocal mics are put back on the stands, and the pastor begins walking up front. He's saying something, but you can't tell because he's facing the wrong way.

As he turns toward the congregation his lapel mic suddenly pops on, a bit loud, but easily adjusted. Maybe. The pastor begins reading from Scripture, his Bible laid open on that super-cool-looking pulpit. Gotta get one of those for your church, or at least your living room. Sounds kinda boomy in the room, but it's a fairly large place. Every time he looks down it gets really loud, pops from time to time, then returns to normal when he looks up. The tone changes, too, but it's doubtful anyone else will notice. What *you* are noticing right now is that the hum is back with a vengeance.

Jonathan has started recording the pastor's mic feed on their computer. He must be running it from the mix bus, since no other auxes are being used—yep, you can see a splitter cable with one side going straight to the audio interface. Won't be the first church to do it this way...wait, what happened to the Pastor's mic? It just stopped. There it goes, no, stopped again. Static. Seems to follow where he walks around on stage.

"How're your recordings turning out? Do you offer them as downloads?"

"Yeah, they're fine. Have to turn the volume up a bit, and the vocals could be a little louder, but people seem to enjoy them. They can also order CDs for $5 or something."

Sunday, May 29...The service ends...mercifully

The last song fades away, the Pastor sends everyone out with the benediction, and all is quiet except for the masses gathering their things and chatting away. A few moments later music suddenly fills the room, and conversation ramps up again so they can hear themselves. After packing up mics and other stuff, Jonathan reaches over and turns off the equipment in the booth... *BOOOMMM...*

"Hey, it's really great of you to be here today. I gotta go home and get some sleep—can you send me some notes or something? Maybe we can grab some coffee next week? Sorry about that—talk to you soon."

Notice anything wrong? Know how to fix everything? Work through the book, then come back to the story again to see if you can figure it all out. I've got the answers at the back (where else?). Just don't cheat...God is always watching, you know.

LEVEL 1

Getting started – the big picture

The Big Picture for running sound at church is simply helping people in the room hear what's going on so they can worship. That's it. Here's your job description:

- Capture sound sources on stage with microphones.
- Make sure they sound good.
- Blend everything together for the congregation to hear.
- Make sure the people on stage can hear what they need to perform.

A sound system includes microphones, cabling, a mixing console, amplifiers, stage monitors, and main loudspeakers. When you talk into a microphone on stage, that signal is sent through cables to the mixing console where you adjust its level (volume), perhaps change its tone (equalizer), and then send it to your stage monitor system and the main speakers in the sanctuary. Here's a diagram to help you visualize this:

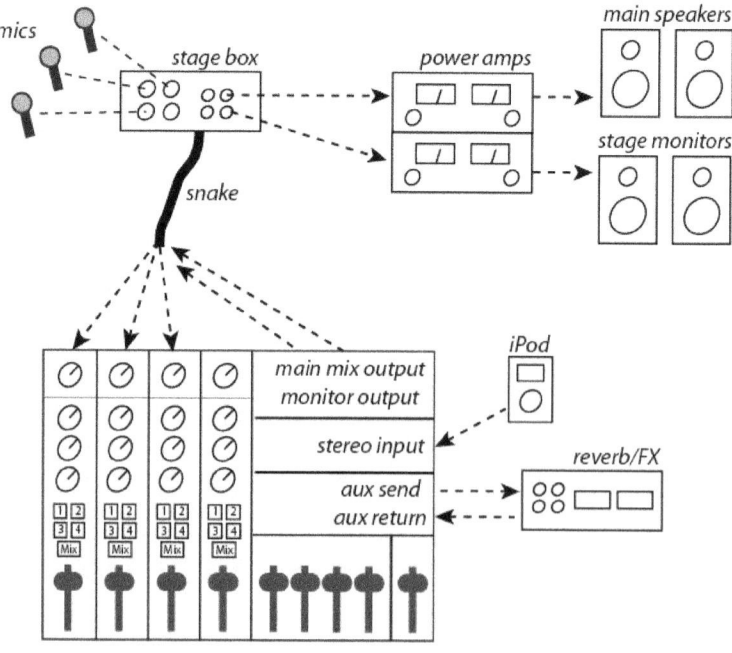

Let's break down the primary operations involved in running sound for your services. Getting a mental picture of where everything goes is really important; otherwise you're just memorizing which buttons to push and you won't be able to figure out new situations and solve problems. I'll keep adding more detail and options as we move along, but you should understand these concepts before diving too deep.

From microphones to final mix

Your main job is to capture sounds on stage with microphones, blend them together, and send this out of your main speakers for the congregation. Here's what's involved:

Microphones are plugged into a connector panel on stage; from there all the mic channels are combined into a single cable called a *snake*, which breaks out again into individual channels at the console. Turn up the *microphone preamplifiers*; the mic preamp increases the audio signal level, then passes it down to the *channel fader*. Turn up the fader and press the *mix* assignment switch, which routes the mic signal over to the *main mix bus*. The mix bus combines all your microphones, creating your sanctuary mix, which is then routed back to the stage via the snake to be plugged into the main power amplifiers. From here it finally goes to the main speakers. You probably know that to create a good blend of your singers and musicians, you adjust the channel faders up and down. There's more to it of course, so we'll fine-tune this later.

The next diagram follows our signal as it flows from the mics, through the console, and back to the amps and main speakers. Note that most of the console doesn't have anything to do with this—you only need to find these highlighted controls on your board. Once you understand what's relevant for a particular task, it makes learning how to use a console much easier.

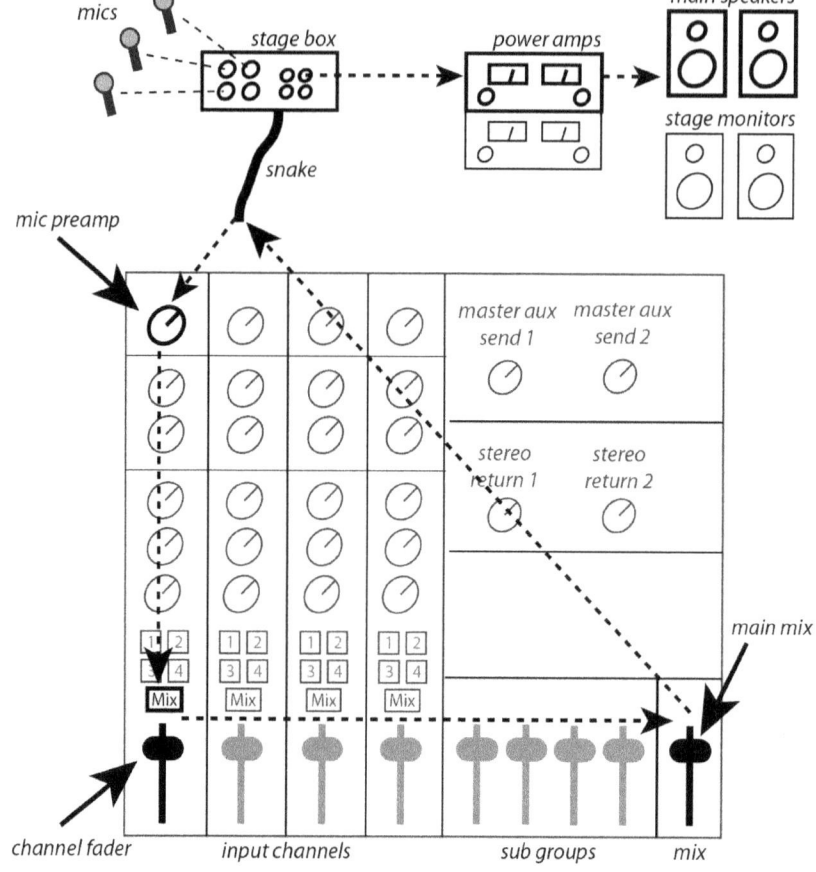

Signal flow from microphones to sanctuary mix

*Microphone > Snake > Console channel mic preamp > Channel fader > Mix bus
assignment > Main mix bus > Snake > Power amp > Main speakers*

Stage monitor mix

Your musicians and vocalists need to hear everything that's being played and sung.
Even though they may be able to hear some of this through the main speakers,
timing and intonation will be an issue due to acoustics and distance around the
room. If a guitar player is primarily hearing the drums after it bounces off the back
wall, he'll be behind the beat. It's hard for people to recognize what's causing this,
and so everyone blames the poor guitar dude. A separate mix needs to be set up that
feeds monitor speakers on stage, headphones, or in-ear systems.

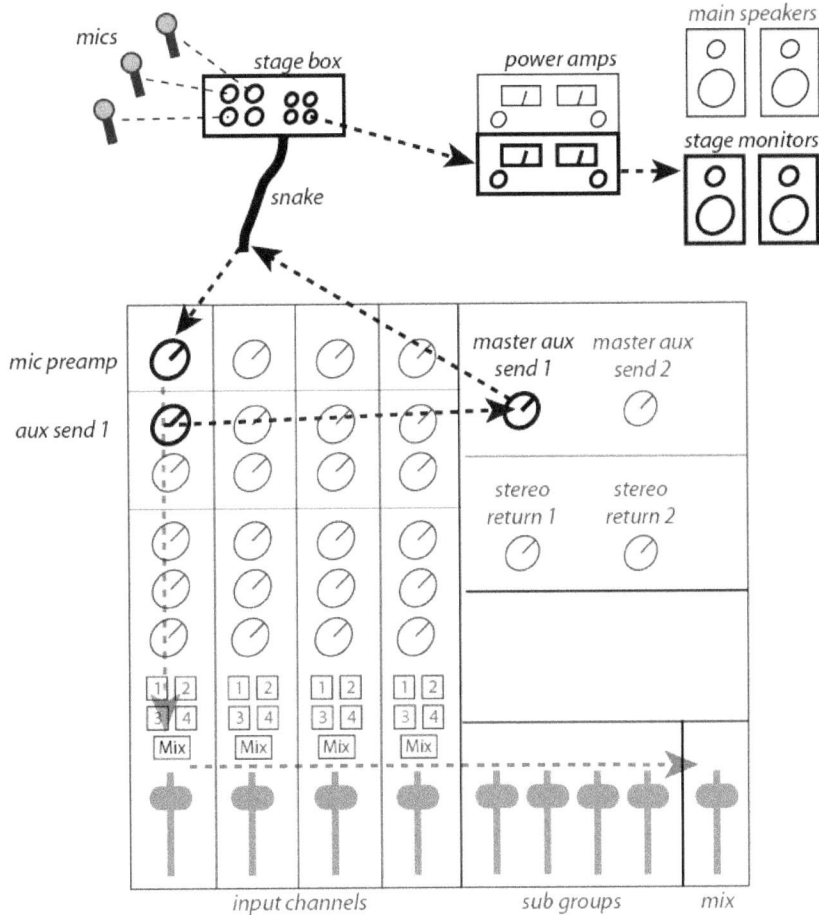

Signal flow for a stage monitor mix

Microphone > Snake > Console channel mic preamp > Channel aux send >
Master aux send > Snake > Power amp > Stage monitors

The same microphone signals coming into the console are used for creating a monitor mix. But, we don't want to send the worship team the sanctuary mix because they will need a different blend of everything. So we grab a copy of each mic signal in the channels and send them to a separate control called an *auxiliary send*. All channel aux sends are combined at the corresponding master aux send (aux send 1, aux send 2); this mix is what you send back to your stage monitors. If the musicians need more cow bell in the monitors, simply turn up the aux send on the cowbell channel, not the fader which affects the sanctuary mix. We'll do more with this later, including examining some different technologies for creating monitor mixes.

These diagrams outline *signal flow*—showing where signals actually go as you set up a mix. Along the way you'll have to turn mics up and down, adjust monitor mixes, add reverb, insert processors such as compressors and noise gates, and figure out problems as they come up. To do this you must understand how everything is connected and routed. Review these diagrams as you read through the upcoming sections; it will help immensely. Even if you're running a digital console, these diagrams show what's happening behind the menus, buttons and touch screens. I'll provide more detail as we get to specific operations, but for now let's get sound up and running for a service.

Level 2

Setting up for a service

It's really early Sunday morning and you've got to set up before the rest of the team arrives. Even if you've been running sound for awhile, take time to go through this section. Most folks have simply memorized specific buttons and switches they push or turn, but that's not good enough.

Power up

First, move your big gulp coffee mug away from the console. Turn everything on, but you need to do it in a certain order. Power up everything except your amplifiers, then fire up the amps. Turn things off in reverse order, so amps go down first. If the amps are running and you then power cycle the other gear, you might get a loud pop through the speakers. This could damage something, so remember this:

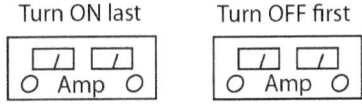

On Stage

Setting up microphones

I'll spend more time later discussing options and issues for microphones, but for now you just want to get signals.

Check each microphone stand and make sure everything is tight—adjustable boom arms, the clutch where you raise and lower it, and the base. I've found many a stand where the boom was about to fall off with a mic on it, so be safe.

Take one mic at a time out of the case. Most of the time it's easier and safer to attach the mic clip to the stand first. Screw this on, then mount the microphone to the clip. You don't want to spin the mic around while trying to screw it onto the stand— the goal is to not drop the microphone. To adjust the stand, first loosen the knobs and retighten them when you're done.

Point it in the right direction. Every mic has an on-axis point which should face the sound source you're miking. Sometimes this is obvious, other times not so much.

For example, long thin mics like the trusty Shure SM57 and 58 are pointed with the end of the grill facing the source. Some mics have a square(ish) shaped grill with two sides. One of these sides is the front—look for the manufacturer's nameplate or use your ears to make sure. When a sound is picked up away from the on-axis direction of the microphone, it will have an unnatural or distant sound quality, which generally doesn't sound very good.

Check the switches on the microphone. If your mic has switches, it will be one or more of the following:

- *Polar pattern select.* You generally want the mic to pick up sounds from the front only, not from the back or sides. Set the polar pattern switch to the heart-shaped symbol; this represents a *cardioid* pattern and means the mic will pick up sounds mostly from the front.

- *Attenuation pad.* This reduces incoming signal level in case your sound source is really loud and is overloading (distorting) the mic. Ever put your ear one inch away from a snare drum? For now, leave it off (0dB).

- *Low-cut filter.* This attenuates (reduces) low frequency sounds such as rumble, vocal pops, and trucks driving by. All consoles have low-cut filters on each channel, so turn those on instead and leave these off. The filter symbol looks like a division sign, so set the switch to the flat position.

Grab a mic cable and connect it to the mic, making sure you hear a *click* that indicates it's locked in. Hold the mic while you're doing this so it doesn't get pushed out of the clip. Microphone cables have different connectors at each end. The male end with three pins gets plugged into the snake panel on stage (audio follows the direction of the pins). The other end has three holes and goes into the microphone. See the little switch on the connector? Make sure you push this when unplugging the cable. You don't need to wrap the cable around the stand like a mummy. Just do a couple wraps around the stand so the cable doesn't hang out and trip someone. Once it reaches the floor, run the cable to your stage snake box in a way that minimizes getting stepped on by people—this damages the tiny wires inside the cable, which is a bad thing.

Electronic instruments such as keyboards, guitars, and bass can be fed directly to the console without microphones by plugging them into a *direct box*. Connect a standard guitar cable into the box's 1/4" input, then connect a mic cable from the XLR (3-pin) output to your snake's stage panel. At that point it operates as if you had a mic plugged into the console. Direct boxes (DI) also have a second 1/4" jack which outputs a copy of the instrument signal. If needed, connect another guitar cable from this jack to a guitar amp on stage where it can be used as a local monitor.

Be careful:

- Don't leave mics on the floor. You just might be the loser who steps on one of them.

- Don't connect or disconnect microphones or direct boxes when the console channels are on. You'll get a mighty crack and pop through the system. Same goes for moving a mic—mute it on the console channel before touching it.

- Make sure your mic cables aren't a trip hazard. Go straight down the mic stand to the floor, then on the floor direct to the panel. Run cables together, not scattered around the floor, to help keep people from stepping on them. If needed, lay down a few strips of *gaffers tape* (no, duct tape is not the same thing—don't use it) to keep it tied down to the floor.

- Make sure mics, cables, and stands are not in the musicians' way; give them the room they need.

At the Console

Setting your main mix

Now let's get the console (mixer) up and running. All consoles essentially do the same thing, but they may be laid out a bit differently and use different terminology. Digital consoles can be particularly daunting, like you've stepped into NASA mission control, but the principles are the same. Follow these steps while looking at your own mixer so you can see how things match up.

All consoles have a number of *input channels* where the microphones are plugged in. These are numbered 1 through 16 (or 24, 32, etc) and start on the left side of the mixer. On the input channels where you have microphones connected, do this:

1. *Condenser* microphones require *phantom power* to operate, so you'll have to turn on the "+48" switch near the mic pre. What's a condenser mic look like? There's no obvious clue, so look at the microphone manual or, if the mic doesn't seem to be working, try turning this on. By the way, some direct boxes also require phantom power; they're called *active* DIs.

2. Select the *main mix bus* switch on each channel. It will be labeled "mix", "L-R", or perhaps "main". This switch sends the incoming microphone signal over to the main mix fader in the master section of the console. If you don't push this, the signal stops in the channel and you won't hear anything.

3. Turn up each channel *fader*. Find where it says "U" or "0". This is the best starting point, and it provides room to turn it up and down as you fine-tune your mix.

4. Now turn up the *microphone preamplifier* near the top of the channel. It will be labeled something like "trim", "gain", or "mic pre". Don't crank it all the way—try twelve o'clock or so (straight up) to get started. You can fine-tune this later.

5. Make sure the channel is on. Sometimes you have to push an "ON" switch, sometimes you just make sure the "MUTE" switch is not on (mute turns off a channel—so this is where you would kill a mic after they're done singing...or if they're singing rather badly).

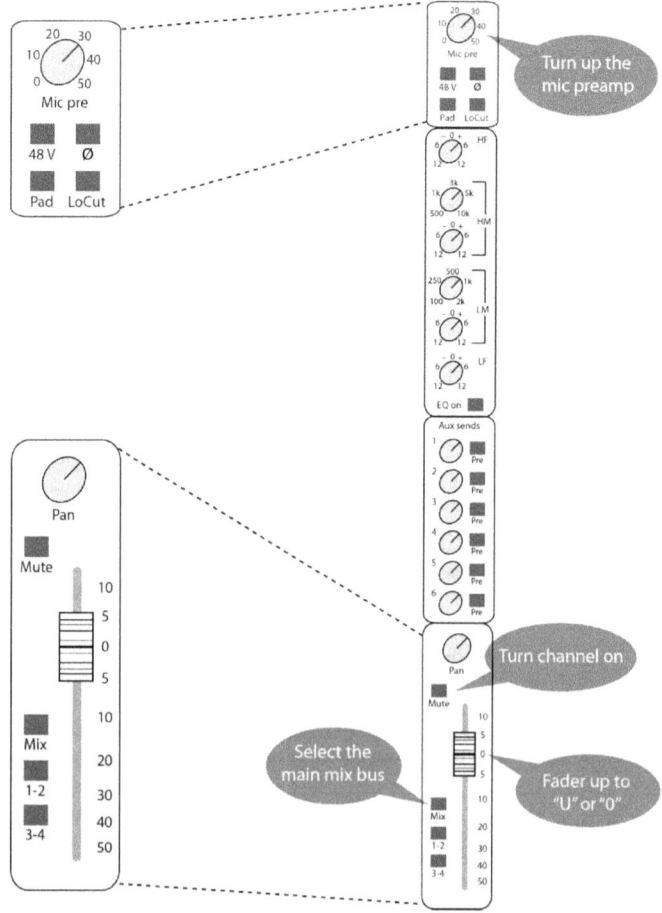

Controls on the channel module to route the mic to the mix bus

Once you've turned on all your mics on the channels, let's get the overall mix out to the main speakers. Look at the master section of your console and you'll find one main mix fader. It's called "main", "mix", or "L-R", and might even be a different color. Turn it up to the "U" or "0". Now, next step...oh, that's all there is. You've taken all your mics, routed them to the main mix bus, and now they're blended together and headed for the amplifiers and speakers via the snake cable. And you thought this was going to be difficult.

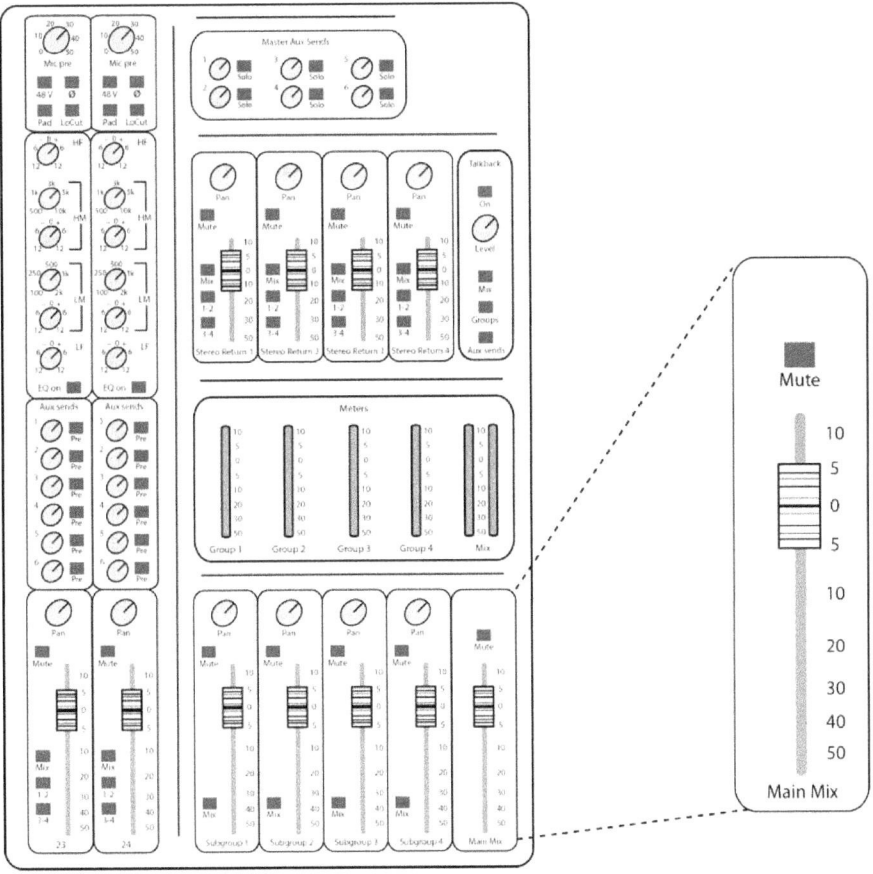

Now all you do is adjust the relative fader levels on your mic channels so you can musically balance the instruments with the vocals and so on. If you've got a fader set near the bottom of the fader path, turn *down* the mic preamp control at the top of the channel so you can bring the fader up into a more workable range. On the other hand, if you've got a fader cranked to the max and you still cannot get enough in the speakers, turn *up* the mic pre control. The idea is to balance the relative levels between the mic preamps, the channel faders, and the master mix fader. Here's the plan:

- Keep the master mix fader at the "U" point.
- Keep the channel faders somewhere close to the "U" point. This gives you room to work as you balance various instruments and vocals in your mix.
- In order to do this, set your mic preamps on each channel wherever you need. There is no magic spot—just get a healthy level so your faders can be optimally positioned.

One more thing to keep in mind. If you change the mic preamp gain at the top of a channel, that will affect all of your stage monitor mixes as well. Why? It's the very first level control on a channel—everything else comes after that, so everything else is affected by it. By contrast, the channel fader is the last level control, so you can change this without affecting your monitor mix (depending on the *pre-fader* switch setting, but I'll save that for later). This is an example of how knowledge of signal flow comes in very handy, so review the charts as you go along.

Here are the level controls you use to set and balance signal levels to the main mix:

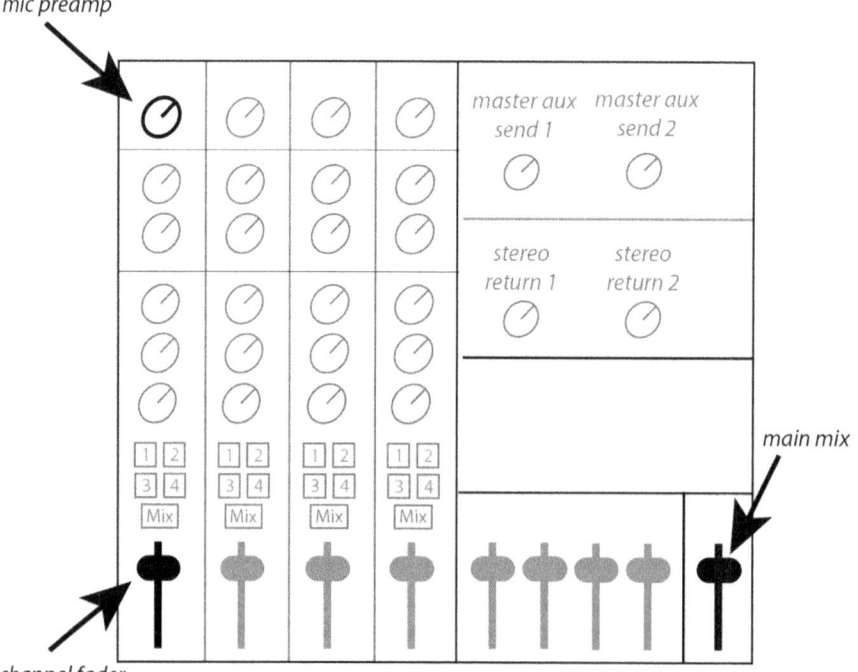

That's it—you should have sound now. If not, try these trouble-shooting steps:

1. Is everything on? (I learned this genius solution from all those tech-support calls over the years.)

2. Is the mix bus on? Some consoles provide an on/off switch for this fader.

3. Are your channels turned on?

4. If certain microphones are not working, mute those channels, turn on the phantom power switch (+48V), and then unmute the channel.

5. Are your mics plugged into the correct mic inputs on stage?

6. Are the musicians actually playing? Maybe they're just messing around and not playing loud enough to get a good signal, or they're unsure about the song. This happens all the time, so don't go chasing them until you see them really kick into gear.

Setting your monitor mix

Go ahead and set up a monitor mix for your worship team on stage. Do this as soon as you can because some instruments, such as bass guitar and keyboard, usually do not have an amplifier up on stage to hear themselves. They rely on the monitor system, so they can't do much until you give them a mix.

Do the following steps while looking at the stage monitor signal flow diagram below:

1. Select the aux send that's connected to your stage monitor amplifiers. We'll use aux send 1 here.

2. Turn up the master aux send 1 in the master section of the console. Aim for *unity*, which may or may not be labeled. If so, stop at "U" or "0". If not, check the manual to find the unity point. If you still can't figure it out, just turn it up to around 1:00 (like on a clockface) for now.

3. On each microphone channel, turn up aux send 1 to around 1:00 or so.

4. Push the button nearby that says *pre*. Some aux sends are permanently set to *pre*, so there is no switch. I'll explain later...for now, just do it and move on.

5. That's it. If the monitor amps are on and everything is connected, they should hear sound.

6. Using the aux sends on your channels, fine-tune the monitor mix to give them what they need. Don't touch the channel faders or you'll mess up the main sanctuary mix. Emphasize piano, drums/percussion, perhaps guitar

and bass. Timing and tuning is what they need, so cranking up the kazoo and other background instruments is not so important and can merely clutter up the mix.

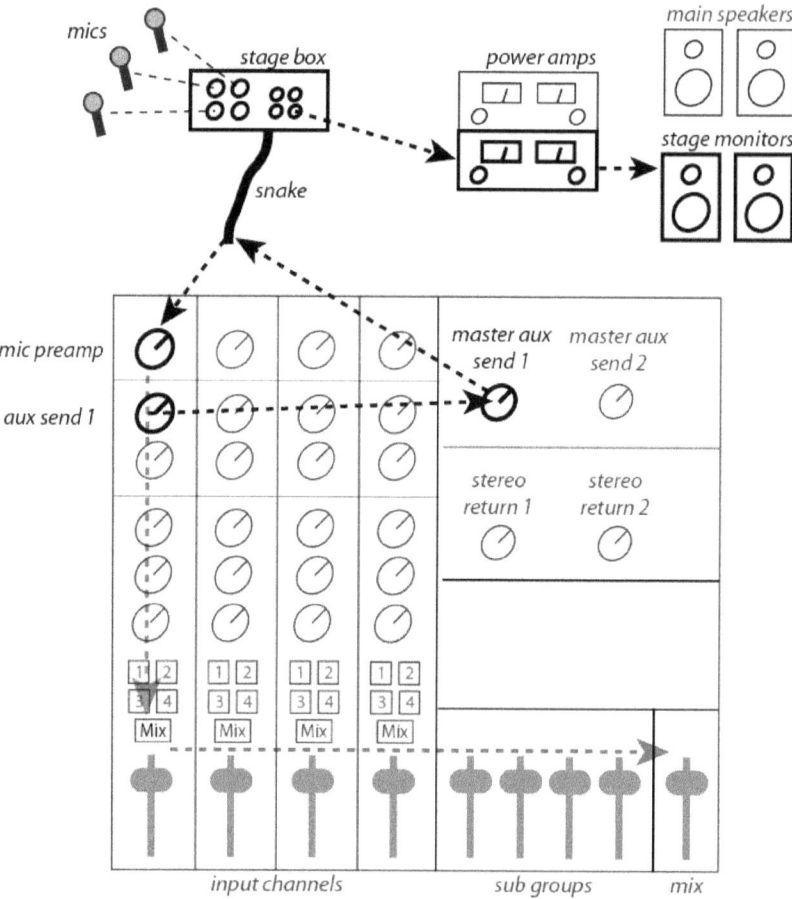

You okay so far? If so, keep going and we'll find ways to make your sanctuary mix absolutely shine.

Level 3

Refining your sounds on stage: Microphones & acoustics

Which mic to use?

Manufacturers intentionally design microphones to sound different. Think of an artist who selects just the right brush from a range of different types on his palette. A sound engineer has a wide selection of microphones to use for various situations. If you browse through an audio gear catalog you'll notice that there are a *lot* of mics available to choose from, but you only need a few to get the job done in a live worship situation. For recording studio work you need to be pickier, but that's not our purpose here.

Dynamic microphones have a rounded, punchy sound that works really well for drums, guitar amps, vocals, and even horns. Many of them are quite inexpensive. *Condenser* mics generally sound brighter and more articulate, and are excellent for strings, horns, piano, drum overheads (cymbals), choirs, and spoken voice situations (you know, the pastor). You can't tell dynamic vs condenser just by looking at them, but they do have certain tonal qualities that help you select what you want. Keep in mind that condenser mics require a special power supply to operate—just make sure the +48V switch is on (near each channel's mic preamp control or on the back of the board). And don't worry, if you leave the switch on and plug a dynamic mic into that channel, it won't hurt it.

Every microphone comes with a chart that looks kinda like an EKG readout. This is a frequency response graph, and it shows how the mic picks up sounds at different frequencies (bass, treble, mid-range). So, if the response line takes a nose dive toward the left side of the chart like the one shown below, that mic will not reproduce low frequencies very well—don't put this mic on your bass guitar cabinet or kick drum. Likewise, if the chart shows a solid high frequency response, it might work well for flutes, cymbals, violins, and piano.

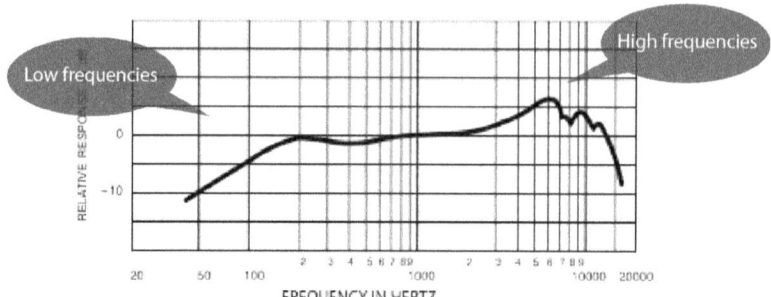

You can also get an idea of whether a mic is good for low frequencies by the size of the mic. You'll notice that some are thin, others pretty large at the grill. All microphones have a membrane (diaphragm) inside that does the work of converting sound waves to an electrical signal. The larger the diaphragm, the better it usually is for low frequency reproduction. Think about the speakers in your home stereo system. The larger speaker cones are called woofers and provide your bass, while the smaller tweeters give you treble. Same here, where the small-diaphragm mics are good for high frequencies, but not as good for lows.

And then there are always exceptions. I mentioned that condenser microphones are great for vocals, but we don't often use them for worship team singers. Use dynamics instead, such as the long-time favorite Shure SM58. Condenser microphones are much more sensitive than dynamics, meaning they pick up sounds better—such as what's coming out of the floor monitors, other instruments around them, and so on. This can translate into earlier feedback issues on stage; you can crank up a dynamic quite a bit before getting feedback. It's a different story for choirs and other ensembles, though, as I'll cover shortly.

I also suggested that microphones with a solid-looking low frequency response chart are best for kick drums. Usually this is true, but sometimes another mic will give you more of what you need. How do you know? Try them, compare, and you and your team make the decision. Nobody's going to arrest you for putting a "wrong" mic on something. The charts and textbook theory can help provide guidance, but ultimately it's up to what sounds good to you. Go look at the mic manufacturer's website for suggestions. They're pretty good at explaining what certain mics were designed for and how to use them, so if they say it's great for kick, well then put it on the kick and experiment with placement. I'll provide some specific ideas for mic selection in just a moment, but here are a few more issues involved in getting good sounds from your mics.

Which way do I point the mic?

Every mic has a front, or *on-axis* point which should face the sound source you're miking. This is never labeled as such, and sometimes you really can't tell just looking at it. Often the manufacturer's nameplate or logo is on the front, but even this may not help. Set up the mic and listen at the console. Does it sound really distant or unnatural? If so, it's backwards, so turn it around. It should sound close and full (unless, of course, the mic is located a long way from the sound source).

Most mics pick up sounds from one direction only (front), although that's somewhat misleading because they always get some amount of sound from the sides. This is called *cardioid* or *uni-directional*. There are other *polar patterns* (direction of pickup) such as *bi-directional* that pick up on both sides of the mic, and *omni-directional* that picks up everywhere. Sometimes the directional pickup pattern is permanent, but some mics allow you to select different polar patterns with a switch. Which do you use? For singers standing in front of stage monitors, you need to use cardioid mics so the pickup zone of the microphones will be facing the vocalists and away from the monitors, thereby reducing the chance of feedback.

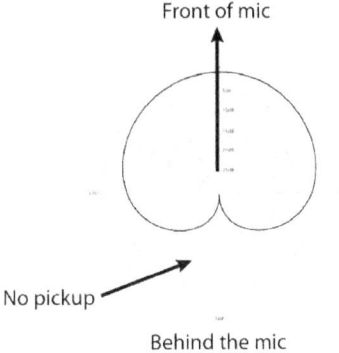

If you want to capture the richness of congregational singing by hanging mics over the seating area, an omnidirectional is ideal because it'll pick up everything in that part of the room.

How close do I put it?

Hmm...depends. When miking individual instruments or vocals, use a single mic placed fairly close. This may be a dynamic or condenser, depending on the instrument and desired sound. Keep it close so the mic focuses on that particular source without capturing too much of what's going on around it. Microphones are pretty stupid; humans can tune out extraneous sounds around us, but mics can't. The closer they are, the better they will capture just that source.

Dynamic microphones in particular need to be pretty close to get a decent sound level, so vocalists should hold it close to their mouth. Mic an acoustic guitar about eight inches away, and for a guitar amp go maybe an inch or so. Snare and mounted tom mics should go really close, say two or three inches from the head, a bit farther for floor toms. If you're miking a choir, you've got to be back far enough to capture a larger group, so we're talking several feet using condenser mics. Same goes for a horn or strings section (violins, cello). The farther away from a sound source, the more room noise and other sounds you'll capture. You'll hear more reverberation from the room, which can be good or bad, depending on what you want. You'll also increase your  chances for feedback since you have to turn up a distantly-placed mic higher. Typically for live sound in church you don't want lots of distant-sounding mics, so keep them closer as long as you're getting a balance of the entire source.

Audio example 1: Close vs distant miking

The closer you are to a source the more you get just that sound—but if you're too close it starts to sound boomy and unnatural. It takes some space for an instrument to radiate a full sound. You'll also get a boost in the bass, so if it sounds pretty boomy try backing off a bit. Shoot for 6-12 inches if possible and listen to the result. Horns need a little more space, so try a foot or two. If you're too tight to the choir, you might emphasize that particular off-key tenor who tends to sing really loud. The example below starts fairly boomy and muddy, then clears up as we move the mic back.

Audio example 2: Proximity effect—mic too close, then pulled away

I'll show you specific miking examples later, but the main thing is to understand the issues involved in close versus distant miking, experiment a bit, and follow your ears.

Acoustic issues for miking

The area around your instruments and microphones can play a significant role in the sounds you're getting. When a mic is placed in front of an acoustic guitar, it captures the sound directly from the instrument, but it also picks up some guitar sound that reflects back from nearby surfaces. When these reflections come from farther away it sounds like reverb, which you might like. However, sound bouncing back from nearby walls, floors, or other hard surfaces is destructive to the tone of the instrument. The reflected sound takes a bit longer to arrive at the mic, so you've got two copies of the guitar sound coming into the microphone at slightly different times.

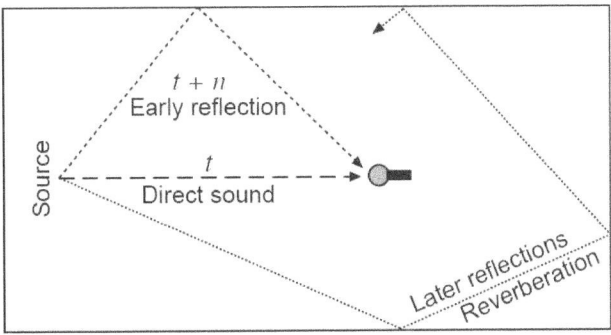

Audio example 3: Phasing (good, then bad)

The result is known as *phasing*, or phase cancellation, and it produces a change in the guitar tone which you probably don't want. It'll sound thinner, somewhat hollow, and you'll lose the richness of your sound. Try to avoid it with these tips:

Try not to place musicians and microphones close to hard surfaces. Either cover the area around them with something soft and fuzzy, angle the surface away from the mic, or move away from it.

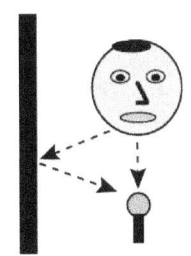

If you use a drum cage, place foam or fiberglass acoustic panels along the inside of the cage panels; this reduces reflections that interfere with the drum mics around the kit.

Mic a grand piano with a PZM mic taped to the bottom of the lid, which can be closed or on short-stick. These mics look like flat plates with a small bump in the center. If you raise the lid and place regular mics in close, you'll get phase reflections from the lid, negatively affecting your sound.

Use only a single mic on each source when possible; two mics facing the same sound source also invite phase effects if they are at different distances from the source.

If your pastor or worship leader uses a pulpit or podium up front, the voice travels directly to the mic as well as to the hard surface of the pulpit, which reflects it back into the mic a split second later. Put a mat or cloth on the surface to reduce these reflections. If you can change the tilt of the top surface, angle it more so reflected sound is focused away from the microphone.

Audio example 4: Podium phasing

Music stands are one of the worst culprits because they're usually metal and are located directly behind the microphone. Sound hits the stand and reflects back into the mic, negatively interfering with the sound of the voice. Try a cloth or mat on the stand, lower the stand, increase the tilt angle, or move it off to the side a bit. Just don't incur the wrath of singers...not a pretty sight...

How many mics should I use?

As few as possible. The more mics you run, the greater the issues of acoustic phasing and feedback. Use only as many as needed to cover the sound source. Here are some examples:

Acoustic guitar and other small(ish) instruments: one mic is sufficient. I like to run stereo mics in the recording studio, but it's not necessary for live sound.

Grand piano: Bigger instruments may need a couple mics simply to cover the larger area where they produce sound, but for piano in live services I use just one PZM taped to the underside of the lid. Listen and move it around a bit to get the best balance of the entire piano sound.

Drum sets: I sometimes use two overhead condenser mics to cover the entire kit along with individual mics on each drum. Some engineers use only the drum mics without overheads, the reason being that in smaller rooms you can already hear the cymbals loud enough. The challenge with using multiple mics is that it practically begs for phase problems to show up since you have several microphones so close together. As you turn on each drum mic, put on headphones and listen to the sound of the entire kit and if it suddenly sounds a bit hollow and thin, you've got a phase issue between a couple of mics. Move them around slightly, changing the distance between them and the drums. Also try flipping the phase switch on one of the drum mic channels and see which position gives you a bigger sound. Your goal is to get a full and natural sound, so experiment until you get it as best you can.

Overhead drum miking using condenser mics in an XY stereo configuration

Audio example 5: Drum miking with phase issues (bad, good, then bad)

Choirs: If it's a small group use one condenser mic. Larger ensembles will require at least two mics, more for really big choirs. Space the mics out several feet from each other and follow the Golden Rule for multiple mics (see below).

Pulpit/speaker: If your pastor uses a lavalier or head-worn mic and you have a mounted mic on the pulpit/podium, do not turn both of them on. Use only one at a time or you'll surely get evil phasing.

The Golden Rule for miking

If you use more than one microphone on a source, follow the long-established 3:1 rule which suggests placing the second mic 3 times as far away from the first mic as the distance between the first mic and the source. Huh? Here's a diagram to make it easier:

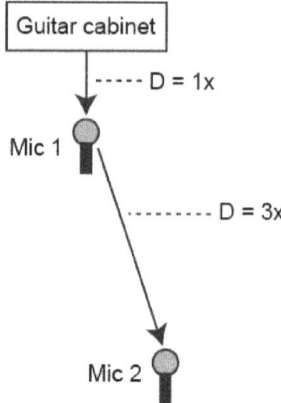

Once beyond that 3x distance the mic signals don't interfere as much acoustically, but within this zone you'll get undesirable phase-induced tonal changes. If you want to experiment, try the second mic at various distances and listen closely through headphones. It might take you a while to pick up on what's happening, but when you begin to hear it you'll be able to fine-tune your mic placements to keep this at a minimum.

Audio example 6: Phasing due to 3:1 violation (good, then bad)

Wireless or wired?

Wireless microphones, besides being much more expensive, are subject to all sorts of issues such as outside interference (RF), bad reception in the room, and batteries that must be replaced. Use a wireless mic only when it's more convenient than a corded mic, such as for the pastor, other people speaking up front, and perhaps your worship team vocalists. Otherwise stick to a corded microphone. You'll have better sound quality, reliability, and save a few bucks as well.

If you're having reception issues such as pops, clicks, and drop-outs, try a new battery, then try moving the antenna on the receiver. You may have to raise the receiver higher to provide a better line-of-sight communication—all those congregants standing in front of you absorb radio energy (we're mostly a big bag of water, after all). Also make sure there is no metal around, such as cabinets or ducts along the ceiling; this can wreak havoc with wireless signals. Beyond that, learn to use The Force. My wife always asks why her cell phone calls get dropped. The answer? It's wireless.

Mic selection and placement ideas

There are lots of different opinions on microphone choices and scads of books and websites to give you plenty of ideas. Here are just a few to get you started for instruments most commonly used for worship teams, along with a few basic concepts that guide mic technique no matter what you're working with.

Groups

Use at least one condenser placed a few-to-several feet away. Dynamics are no good from a distance. Keep the mic back a little to let the individual sounds in the group acoustically mesh together. Too close to a group singles out a particular instrument or voice it's pointed at. Place the mic a bit higher than the group, facing down towards them. This helps prevent the front and center folks from firing straight into the mic and gives you a better blended sound.

Before you start fiddling too much with placing mics, work with the group to help them balance themselves. Arrange them so you don't have a particularly loud instrument right in front. Quieter instruments, such as clarinets, can move toward the center of the group closer to where the mic will pick up strongest. This is true for any type of ensemble—if they get their own balance and sound just right, it's much easier for you to capture it with microphones.

Drum set

Generally dynamic mics are best for drums, with condensers used for cymbals and hi-hat because they are brighter and clearer in the high frequency range. Make sure your mics are out of the way of flying drum sticks; it's usually a compromise between finding a good spot for sound, yet staying out of the drummer's way.

Overheads/cymbals: A pair of condensers over a drum set is ideal for capturing not only the cymbals, but most of the entire kit. Always use an overhead pair in either XY, as shown here in the photo, or a small spaced pair configuration where each mic is located off to the sides of the kit aimed downward toward a cymbal. Locate them two or three feet above the cymbals—too

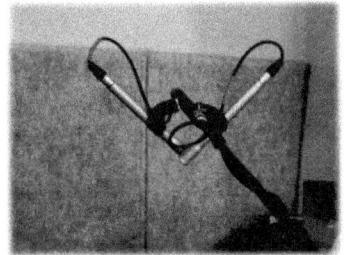

close and they'll sound harsh, too far and you'll get a bunch of other sounds. A hat mic is often not needed since the overheads will get it, but if you want a more direct sound you can place a condenser over to the side pointed downward toward the middle of the top cymbal. Don't face it directly into the side where the two cymbals clash together; if you don't believe me, go stick your face in that spot and wait for it.

Kick drum, floor tom: Use a large diaphragm dynamic mic like the Shure Beta 52, Heil PR48, or the Sennheiser e902. These are good for rich, punchy low frequencies. For a kick drum, remove the front head (or at least have a hole in it) and place the mic inside the shell facing toward the drummer. For a sharper attack, place it a few inches from the back head and aim it where the beater hits. To get a deeper, punchier sound pull the mic back into the shell at least halfway. Experiment with different placements and you'll get very different sounds. Stuffing blankets, sweatshirts, or jelly doughnuts up against the head inside the drum dampens the resonance, giving a more controlled sound. For floor toms, locate the mic a few inches inside the rim and around 3-4 inches high; too close and it'll sound boomy and unnatural, too far and it begins picking up other sounds around it.

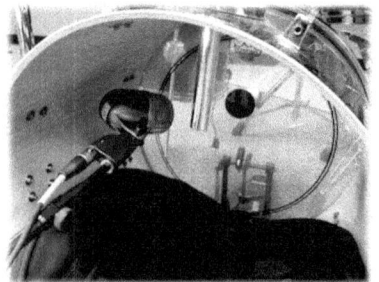 

Snare drum: The Shure SM57 is a long-standing favorite. You want a mic that can capture that crisp attack of a snare, and the 57 has a bump in the high-mid frequency range that's perfect for this. Place it over the edge of the snare, maybe two inches inside the rim, and a couple inches high.

Mid-toms: Similar to the snare, you want to emphasize the attack sound of the tom, where the drumstick whacks it. The 57 works fine, but it begins losing lower frequencies that larger toms have. For a fuller tom sound try the Sennheiser 421, Heil PR28, or experiment with other dynamic mics. Place them like we did for the snare, maybe a little higher.

If you don't have a bunch of mics to try out, don't worry. There are affordable, complete drum miking packages available that include microphones for each part of the kit. Then all you have to do is experiment with placement options to fine-tune everything. I finally bought one of these some years back, and when we switched from a single, pathetic mic suspended over the entire drumset to a full-mic setup, everyone went "wooowwww". Indeed. And the Lord rested on the next day.

If you mic all the drums and don't have enough snake channels available, consider running a separate small mixer on stage to create a drum sub-mix. All of your drum mics go to this mixer which outputs a mono or stereo mix to plug into the stage snake. You can't control individual drum balance levels from the main console, but you save a bunch of channels that you might need for other stuff.

Free tip: Check out the wear patterns on the drum heads. If they're tightly contained in a small spot, you've got a good drummer who probably won't bash your mics. If you see stick marks all over the place, watch out and pull the mics back.

Piano

Condenser mics are ideal for piano due to its very complex sound and extensive frequency range. One popular option is to tape a *PZM* (*boundary mic*) to the underside of the lid. These feature a condenser mic capsule mounted on a flat plate. The small end of the housing on top is where it picks up sound, but you don't really have to point it directly at anything. Just make sure the top side of the plate with the housing is facing toward the sound. Once I found a church team who mounted these types of mics on stands and faced them away from the sound source, not knowing this was actually the back. The other problem here was that boundary mics are designed to mount on a large surface, such as a piano lid, wall, or floor. Don't put them up on a mic stand.

We usually don't place regular microphones on stands into or in front of the piano for live services. You'll get everything else on stage leaking into the mics along with various levels of phasing from the piano lid. While you can purchase specific piano miking systems, they're expensive. Get a boundary model and tape it with gaffers tape (not duct tape, which leaves an awful sticky residue). If I'm using only one PZM, I'll position it over the high strings for more clarity and presence in the mix, rather than the heavier low end.

Acoustic guitar

Many acoustic guitars feature an electronic pickup with a 1/4" jack. This is the best situation since all you have to do is plug it into a direct box; you get a really clean sound that completely avoids leakage issues. However, this sounds different from the natural acoustic sound of the instrument. If your team is not happy with the sound, or if the guitar doesn't have a pickup, put a mic about 6-8" away in front of the instrument, pointing between the sound hole and the upper fret board. If it's too close toward the sound hole it'll sound boomy, and too far up the neck sounds stringy with little tone. Experiment toward the middle and move it around a bit.

If you use a dynamic mic, it needs to be fairly close to get a good level. Otherwise you have to really crank the mic preamplifier on the console adding lots of noise. But, if you get too close it'll sound boomy no matter where you point it, so find the sweet spot and nail your guitarist and her instrument into position so she doesn't move around.

In the studio, engineers use condenser microphones on acoustic guitar because they give you a clear, bright, articulated sound. You can try this, but again be aware of the greater mic sensitivity that'll capture other stuff around it as well as increase your chances for feedback. For live situations it might cause more issues than it's worth.

Electric guitar

There are two options for electric guitar—straight from the instrument into a direct box or placing a mic in front of a guitar amp. The amp is usually a pain because they're *loud*. Some of this is probably due to hearing loss after too many Stairway to Heaven jam sessions, but it's also because the cool sound you want from an electric guitar requires cranking it up to eleven and beyond. Of course this is bad as it will leak into all the other mics; it also wrecks your main mix since you're hearing the amp from the stage competing with the console mix coming through the main speakers. If they insist on using the amp, try to contain it by pointing it away from the congregation, preferably angled up toward the musician so they get a direct shot in the face. This way it doesn't have to be turned up quite so much. Other options include putting the amp in an isolation box that you can buy or build yourself, or even shove it in a back room.

So, what mics are good for guitar amps? The venerable SM57 has been a solid go-to mic for decades (because it's cheap and has a bump in the upper-middle frequency response that helps emphasize the presence and crunch sound of the amp). You can also find mics specifically designed for guitar amps, such as the Sennheiser e609. Place the mic in front of the amp grill an inch or two away, off-center of the speaker cone inside. Experiment by moving the mic closer or farther back and pointing toward various areas of the speaker cone.

By the way, we normally use dynamic mics for guitar amps, live and in the studio. Why? They simply sound better for what guitar dudes are looking for. Personal confession—I once tried to put a very expensive condenser microphone on a guitar amp for a recording session. My logic was that I had this really nice, name-brand microphone (spelled N-E-U-M-A-N-N), and I figured it would blow away the cheap 57s that the band used. Boy did I get an earful...and yes, it was awful (I was very young and dumb in those days). Now, having said this, Alan Parsons (engineer for *Dark Side of the Moon*) likes condensers and ribbon mics on amps, so there's a reminder that there are no rules!

Another option to get a great sound without using microphones and loud amps is to use an amp simulator. These devices are completely electronic, no miking required, and get plugged in between the guitar and a direct box input. You can choose different actual amplifier sounds by selecting the various models and effects programmed into the device. Very cool, not very expensive, and makes a huge difference for controlling your stage volume and getting a solid main mix. Sounds like an ideal Christmas present for someone you love.

Bass guitar

This has got to be the easiest instrument on earth to work with. Just plug it into a direct box with a standard 1/4" guitar cable. Remember, they cannot hear anything they play until a monitor mix is running that will feed the bass signal back to their headphones or floor monitor. Now, when they ask you to plug the bass into their monstrous bass amp/cabinet  sitting next to them, say *no*...at all costs. Guitar amps are bad enough since they always leak into other mics on stage and play havoc with your house mix, but bass is especially grievous because low frequencies go *everywhere*. You cannot contain them, and you can't point the amp in any particular direction and do any good. Offer to buy them coffee every week and they usually go away happy.

Want a better sound? Use an *active* direct box. These models use different electronic components that output a more powerful, edgier, punchy sound. Similarly, bass guitars come in both passive and active models. Active basses output a higher signal level and punchier sound that makes a big difference for your mix. Don't forget that active DIs require phantom power (just turn it on at the console channel). For active basses, this requires a 9V battery in the instrument. When the battery runs low you'll have signal problems, so this is one of the first things to try if you're having trouble.

Audio example 7: Bass running through a cheap vs quality DI

Keyboard

Did I say that bass guitars were the easiest things to plug in? They actually run a close second to keyboards. These electronic instruments go straight into a direct box. No amp simulators, no batteries to go dead. Doesn't get any better than that. Look for the *main left* or *mono* output on the back of the instrument. Most of these do not feature built-in speakers, but even if it does you don't want to use those. Connect a DI and run a monitor feed from the console so they can hear what they're playing. Again, buy a good DI, not the $25 specials you find in the grocery store checkout line. One of my musicians once asked me why his keyboard sounded so awesome with his headphones plugged straight into the instrument, but lifeless when he used the monitor system. I took one look and saw the crappy DI sitting on the floor. We swapped in a much nicer, active DI and boy, what a difference! Don't go cheap on your audio gear. It's not worth it.

Instrumental ensembles

This depends on how many and what instruments you're dealing with. We like condensers for their nice articulation and bright sound. They also pick up better from where you need to place them (several feet away), whereas dynamics don't do well at this distance. You really don't want to mic each instrument individually; let the group create their ensemble sound, then you simply capture the magic. So, if you've got a couple players sitting next to each other, a single condenser microphone two or three feet in front of them will work fine. Same goes for a small ensemble, such as a string quartet or brass quintet, but you'll back it away a bit to cover the wider area and avoid getting too much of the players in the middle.

You should modify the exact placement depending on the type of instruments; they radiate sound in different directions. For example, trumpets go straight out the front, so be careful putting a mic directly toward their bells. Clarinets point down to the floor, but don't put a mic down there; point it toward the middle of the instrument. Cellos should be mic'd lower down where the bowing action is, and violins radiate upward. I don't want to get too complicated here, so try this:

Small brass/woodwinds group: Single condenser mic in front, slightly higher than the players and angled down toward them so nobody's playing straight into the mic. Keep it several feet away so it captures everyone as equally as possible.

Small string ensemble: Try the same approach with a single condenser in front. That should be sufficient, but if you have a cello you may end up adding a condenser mic on it since it radiates sound lower toward the floor than the others. Keep this one fairly close to the instrument, slightly off to one side, pointing toward the area where the bow hits the strings. Watch out for the bow, though. Cellists get in the groove and that stick goes crazy.

Cello accent mic

Large orchestra or wind ensemble: This becomes more complex due to the larger size and different instrumental sections to capture. Try to use as few mics as you can get away with due to feedback issues (more mics equals more feedback). So, try a pair of condensers in front of the entire ensemble set in a stereo XY configuration. This technique works well no matter if you're running a mono or stereo mix in the sanctuary due to minimal phasing between the two mics. This should be placed fairly high in front and above them (10-12 feet from floor level, over the conductor) and angled downward facing behind the first couple of rows. The idea is to not get too much of 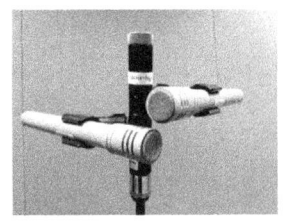 the front row, capturing the entire ensemble sound as it radiates upward and forward. You might also need to place individual mics to pick up instruments or sections that are difficult to hear such as soloists, a piano located off to the side, or percussion in the very back. For these *accent* mic situations, follow the single mic recommendations described earlier.

For larger orchestras in a pretty big room, you might have to do more than the stereo pair out front with one or two accent mics sprinkled in. Here are a few photos of a large choir and orchestra in a fairly big auditorium. Feedback, room acoustics, and the number of mics required are our enemy here, so we keep mics pretty tight on each section to minimize cranking up the mic preamps. This also helps reduce undesirable leakage, such as when the piano and drums get picked up by the choir mics.

Sax & clarinet sharing a mic

Overhead miking for violins, two players per mic

Choir mics placed very close due to leakage and feedback

For all of these scenarios, experiment with spacing and distances so as to capture an accurate, clean image of the ensemble. If the instruments sound too scratchy or close, back the mics off a little by raising them a bit higher or moving farther away. If you're getting too much hall reverb or feedback, then you're too far away.

Of course, if your sanctuary isn't that large you probably don't even need to mic the orchestra. Problem solved.

Vocals

Individual singers, such as vocalists on your worship team, need to have their own microphone. Group miking doesn't work well for these situations. This is usually a Shure SM58, Heil PR20, or something similar, and it needs to be held fairly close to the mouth. This lowers the amount of signal gain required on the console, thereby reducing feedback problems. Show them how to hold the mic right below the mouth with the grill facing upward, not up in front of their face like a pop star. This position is best for sound pickup and avoids those nasty pops and booms that come from "p"s and "b"s. Keep an eye on them as they sing; there's usually quite a bit of mic waving in the air, so encourage them to hold the mic still without killing their enthusiasm.

Head-worn mics position the mic capsule element (the part that picks up sound) to the side of the mouth. These are great because they keep the mic stationary no matter where the singer goes. Same for the pastor; the mic stays consistent as they turn or walk around. If you've used lavalier (lapel)

mics you know what I'm talking about as the sound changes drastically when the pastor turns their head while talking or looks down to read. By the way, these mics sound fine for what we're doing here, but you don't want to use them in a recording session.

Choir

Choirs are mic'd similarly to what we discussed with the instrumental ensembles. A single condenser works fine for smaller ensembles, while many choirs are usually big enough to use two or more. Don't overdo it, though; more mics equals more feedback. Too many close together can also cause tonal problems due to phasing. Remember the 3:1 rule? If a mic is, say, 6 feet from the front of a choir, try to keep the second mic at least 18' from that first mic. So, if you're running 2 or 3 mics to cover a wide choir, keep them apart from each other.

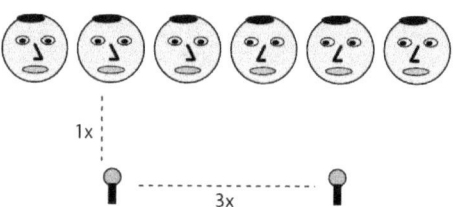

Angle the mics down facing about midway between front and back rows. Remember the on-axis cone where a cardioid mic will pick up sound; try to keep your source within this cone, and also watch that the mic is not in danger of picking up your main speakers. This can be tough when the choir stands near the front of the stage directly under the speakers. In this case, go a bit higher with the mics and point farther down so the rear of the mics face the speakers (the off-axis rejection zone). Don't put mics too far back from the group; you'll have to crank up the levels too much, resulting in feedback and a lot of room sound you don't really want.

Choir mics placed high and pointed downward toward the middle rows

Now, listen and see if you're losing people located between microphones; if so, try pulling the mics a bit closer together to cover the hole or add another mic in the center. Also make sure you are capturing a fairly balanced blend of the ensemble. If people in front of the mic are sticking out, pull it back. You may have to slide the outer folks closer in so they can be heard.

Instead of putting mics on stands in front of the group, you can also hang them from the ceiling. This way you can leave them in place permanently, they're out of the way, and most people won't even notice them. Don't hang a big fat condenser over your choir, though; several manufacturers make specific models for this type of situation. These mics are very small with thin cable and a bracket that angles it toward the source.

Here's an idea...

As much as you can, make the effort to learn about each of the instruments you work with. How do they work? Where does the sound come from? How are they tuned? Ask your musicians to explain their instruments—they'll love the opportunity to show off what they do, and you'll be far more knowledgeable about capturing their sound and troubleshooting problems along the way.

Refining your sounds on the console: Mixing and processing

What's a mix?

Just getting mic signals into the console and out to the main speakers is not enough. You need to build a mix that does the following:

- Musically balances the various instruments and vocals.
- Sounds pleasant to the ear—not harsh, piercing, or muddy.
- Reduces distractions from noise, feedback, and other undesirables.

How do I do all of this?

First thing is to work on your listening skills, meaning you've got to be able to hear what a good musical balance is and how things are supposed to sound. Years ago we had a dear lady who ran sound for us; she was wonderful, but had no idea what she was listening to. Since I played piano during services, I would use secret spy signals to indicate any changes she needed to make. She sat up there in the booth with both eyes on me, waiting for the next update.

Developing your ears takes lots of practice and guidance, especially if you're not an experienced musician. However, don't take this to mean that if you're not "musical" that you have no hope. Nearly anyone can learn what a good mix sounds like, so dig in and see what you can do. Listen to professional recordings of the songs your church uses so you know what you're shooting for. I don't mean play these in the car on the way to church, but sit down and really *listen*. How loud is the snare drum in this mix? Vocals? What should an acoustic guitar sound like? What's important here?

Once you get a handle on this, you then tweak it a bit more for the specific needs of a worship service. When I was in college I ran sound for theme park shows (with the awesome benefit of free coaster rides for the entire season). At first I carefully crafted a fine-tuned, nuanced studio mix, like I'd do for records. It wasn't long before the show producer came over, pushed me out of the way, and cranked up the vocals (with *lots* of reverb). "That's the way you do it!" Ok, lesson learned. The vocals *were* the main feature for a live show. Drowning them in reverb covered up the imperfections of a group of singers trying to sing and dance around on stage.

On top of all this is making sure nothing gets in the way. How many times have you cringed at feedback, waited for someone to turn on the mic, or got a headache from that piercing trumpet that's firing straight into the microphone? Here are the main options and issues involved in tweaking your mix.

Musical balance

All your vocals and instruments should be blended so you hear the more important parts, usually vocals, more prominently, while the band or ensemble fully supports them. You want to hear everything, but in context to their role. In other words, you want to miss it when it's gone, but don't draw attention to it.

Use your faders on each channel to balance all of this. Try not to adjust the mic preamps at the top of each channel, because this will change the monitor mix on stage (since the mic pre is before the auxiliary sends used for the monitor mix). Set your mic preamp levels so that the channel faders are around unity (0 or U), then mix from the faders.

One other thing to keep in mind is that some of your parts will be heard acoustically in the room, meaning you don't have to turn them up as much in the mix. In my church, if you sit along the left side, you'll hear lots of piano and drums along with whatever comes out of the main speakers. This is also where the console/mix position was located for years (wasn't my idea...), so whoever ran sound had to verify what the mix sounded like on the *other* side of the sanctuary. Over there you don't hear as much acoustic piano and drums, so unless we're careful, they'll get more keyboard and vocals and not enough of the other stuff.

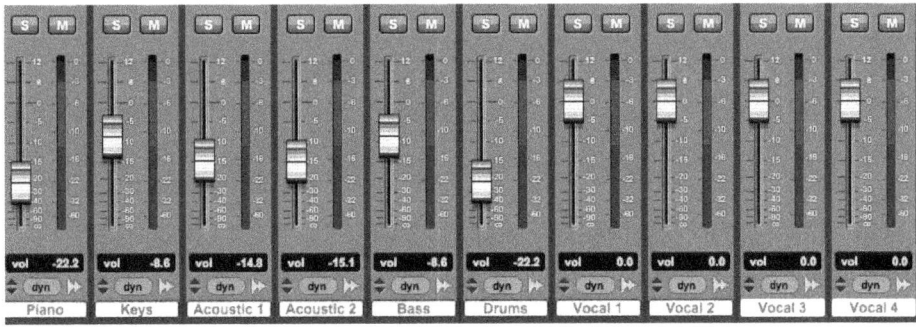

A typical sanctuary mix. Note the softer parts that are cranked pretty high (vocals) and the louder acoustic instruments that don't need as much (piano, drums).

So what exactly are we shooting for here? What *should* it sound like? Let's run through a few examples. First, here's a before and after example that has a balance problem in the mix which is then corrected by changing the fader level for that part.

Audio example 8: Guitar too low in the mix, then better

In this mix, the bass guitar starts too high.

Audio example 9: Bass is too high

Audio example 10: Works better

Your turn now. Listen to the following excerpt and see if you can tell if something's not quite right.

Audio example 11: Something's not quite in its place

Now listen to a slightly different version of the same example.

Audio example 12: Better

What changed? The organ was too loud at first, so the fader was turned down to place it better in the mix. Listen to another one and see what needs fixing.

Audio example 13: Something else is wrong

Audio example 14: That's more like it

This time the drums were too low at first, but fit better in the second mix.

And yet another mix that needs something, but what?

Audio example 15: The vocal is too high

Audio example 16: Now it fits better into the mix

You'll find it's easier to hear things that stick out, but more difficult to notice something that's not quite loud enough unless you're specifically listening for it. The congregation will always hear that loud, shrill soprano, but may never miss an alto that's barely there (except her husband). One more—what's missing in this mix?

Audio example 17: Something's missing here...

The kick drum is completely gone. If you had a tough time hearing these issues, don't feel bad. It's especially difficult when you're not sure what you're looking for. I can listen to a mix several times and find different things going on with each run-through. Keep practicing.

Tone

At some point you've undoubtedly heard (or ran) a mix where a vocal was edgy, the bass a muddy mess, and the pastor's mic sounding like it was spitting static. Along with finding the best mic and placing it in a good position, you can control the tone of a sound using the EQ (equalizer) on each channel of the console. Too bright? Turn down the high frequencies (treble). Too muddy? Turn down the low-mids. Too edgy? Turn down the high-mids. Piano sound a bit dull? Turn up the high-mids a bit for more presence.

Listen to these examples and see if you can hear any tonal problems.

> Audio example 18: How do you like these cymbals?

Ouch. Way too bright, edgy, and painful. The high EQ control was cranked up beyond reason.

The next example has the opposite problem—it's too muddy. You need to clean up various instruments by attenuating the low-mid frequency region around 200-400Hz.

> Audio example 19: A muddy, boomy mix that gets cleaned up

What's going on here?

Our perception of tone quality comes from the unique set of frequencies (sine waves) that are present in any particular sound. When you change these by increasing or decreasing in certain areas you'll alter the tone of your sound. So, a vocal that's too bright needs the high frequencies reduced a bit; this is accomplished by turning down the HF (high frequency) control on the EQ.

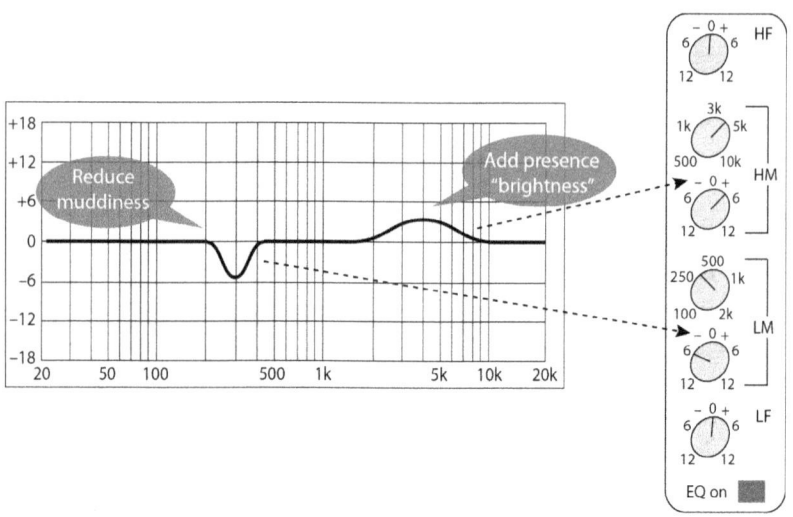

High-mid frequencies are boosted 3dB at 4kHz, resulting in a brighter sound. To reduce muddiness, the low-mid region is attenuated 6dB around 300Hz.

EQs divide our hearing spectrum (20Hz to 20kHz) into regions called *bands*. The EQ shown above has four bands: high, high-mid, low-mid, and low. Your console may be a bit different, but it does the same thing. Note the attenuation in the low mids along with a boost in the high mids. Let's do that to a snare drum and acoustic guitar. The lower dip reduces some cloudiness while the upper boost adds snap and presence, helping the part to cut through in the mix. I'll turn the EQ on partway through so you can hear the difference.

Audio example 20: Snare drum–EQ out, then in

Audio example 21: Acoustic guitar–EQ out, then in

That cloudy sound comes from the fact that all acoustic instruments and voices have what we call a *resonance* region, meaning it will sound a bit muddy in the low-mid range. This comes from too much energy in this band, so we use the low-mid EQ control to reduce it. This is usually the first step I take when dialing in an EQ.

Audio example 22: Muddy bass guitar cleaned up with low-mid attenuation

Audio example 23: Keyboard using low-mid attenuation, EQ off, then on again

These EQ settings may sound thin to you, but in a mix with lots of other stuff this is how things get bogged down. We often over-EQ certain things to help it fit into the overall sound.

This example takes the overhead mics on a drumset, which sound a bit dull for the cymbals, and boosts the high frequency range a bit to make them brighter and crisper.

Audio example 24: Drumset overheads–EQ out, in, out, in

Here's an entire drum set without, then with EQ on each mic channel.

Audio example 25: Entire drum set before and after EQ

Again, if you're having a difficult time hearing some of these changes don't give up. It takes a while to get a feel for EQ, so practice and listen. However, remember that some tonal issues can (and should) be fixed by moving or changing a microphone, or even fixing something on the instrument itself. Does the guitar sound muddy? Move the mic back a bit or point it away from the sound hole; maybe it needs new strings. Snare drum ringing a lot? Tune it or place an O-ring around the perimeter. There's usually more than one way to solve a problem. We'll do lots more with EQ later in the book, so stay tuned.

Reverberation (ambiance)

Nobody listens to music in a total vacuum. There's always a surrounding environment that affects what it sounds like; the way sound waves reflect around the room gives us aural clues as to what type of space it is. Bathrooms are very vibrant, highly reflective spaces, as opposed to living rooms that are fairly muted thanks to lots of couches, plants, bookshelves, and people. Gymnasiums are usually the worst for trying to understand what the announcer is saying, again being a very reflective, boomy room. Most sanctuaries fall somewhere in between, so you'll have at least some ambiance for the music to mesh with.

Audio example 26: Dry–no reverb

Audio example 27: Hall reverb

When we close-mic sources, we lose much of the room's ambiance and get a dry sound. This is usually a good thing for live situations as it provides us with more control over each microphone. But, we want some amount of reverberation in our music, especially for vocals, to help them blend better and sound more natural. Ever listen to a close recording of someone singing, or perhaps you actually did go out and listen to the hallway speakers and hear those vocals with no reverb at all? Yikes. It sounds thin and you hear all the imperfections in pitch and timing. Adding a bit of artificial reverb will cover much of this. Your sanctuary will add its own reverberation to your mix (which may or may not be a good thing, depending on the acoustics of the room). You can also add artificial reverb with effects processing. However, use only as necessary; adding reverb to the pastor or other spoken sources will detract from what they are saying, making it more difficult to understand. So, turn off the reverb when people are speaking, back on during the music. Here's how it's done on an analog system; the principle is the same for digital boards.

Pick one of your aux sends to use for reverb (we'll use aux 1 for now). It needs to be a *post-fader* aux (more on this later). On the channels you want to add reverb, turn up aux send 1. Now turn up master aux send 1. The output for this will connect into your reverb unit, then come out of that box back into any stereo aux return (or a spare, empty channel). Look at the diagram to help make sense of all this.

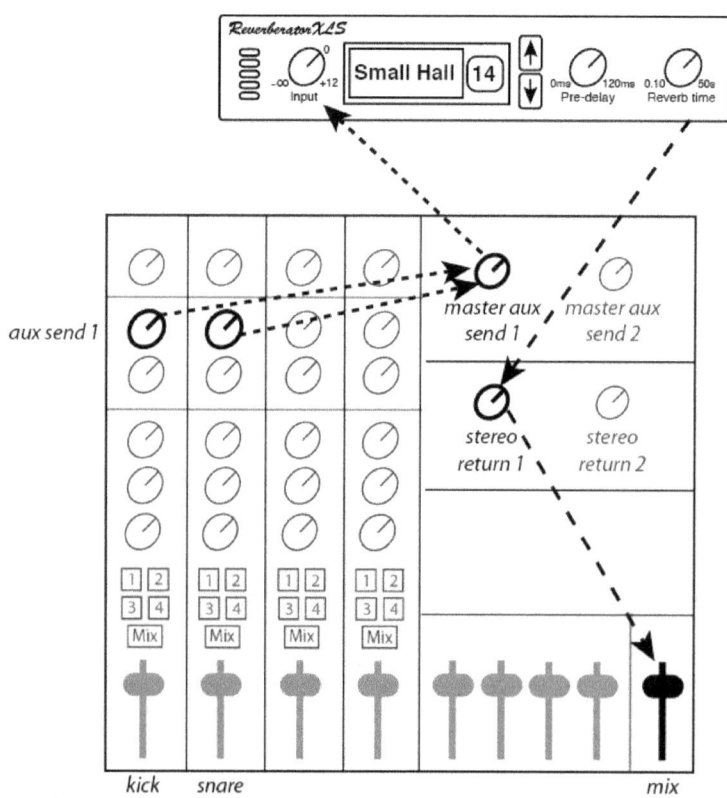

Since we have an aux 1 on all our channels, you can send anything to the reverb unit simply by turning up aux 1 on those channels. To hear the reverb, it has to come back into the console and get routed to the main mix bus. Some consoles have dedicated aux returns (sometimes called *stereo inputs*) that feed the main mix, so connect your reverb outputs into one of these returns. If you have spare channels, you can plug the reverb unit's output into a channel (line input), then route this to the main mix like all the other mic channels. Caution—don't turn up aux send 1 (or whatever aux number you're using) on this channel or you'll get an electronic feedback loop. I'll let you chew on this awhile to see why.

Note that the reverb signal does *not* go back to your mic channels. Lots of folks believe this, and from the surface it sounds like it. But, what's really happening is that you've got your dry mic channels and the reverb sound converging at the mix bus, creating the illusion that you're adding reverb to that vocal on channel 5. Just to clarify, the reverb unit does not output individual sounds with reverb on them; the reverb sound coming out of your device is a combination of *all* the mic channels you sent via the aux send, creating one signal of reverb that needs to go to the mix bus.

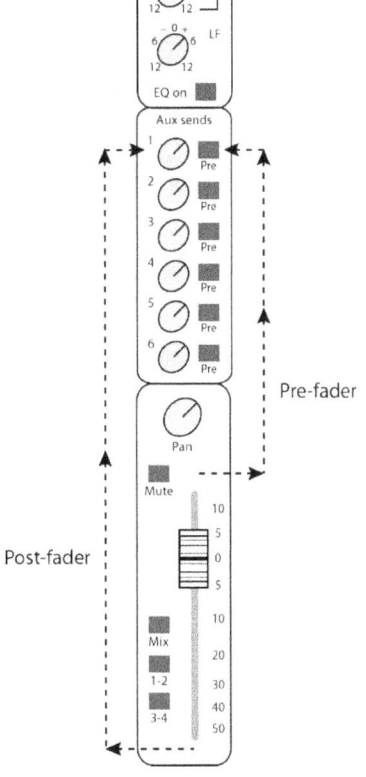

If you're running a digital console, it's got built-in effects so you don't have to connect an external unit. But, the signal flow is the same internally—you turn up channel aux sends, which are routed to your effects engine, then that signal has to return somewhere in the console to feed the mix bus. It's just harder to see than with an analog setup, so the above diagram can help you understand what's going on inside.

Now, back to that pre/post-fader thing. As opposed to monitor mixes which require a pre-fader aux send, we almost always want the aux send to get its signal *after* the fader level for reverb. Why? When you make adjustments to fader levels during a service, you usually want the amount of reverb to match. So, if you set a reverb aux send pre-fader and need to turn down the vocal fader during the service, the main voice sound is lower, but the reverb level stays constant. You'll hear lots of reverb and little vocal. Set the amount of reverb you want with a post-fader send and the verb will follow your faders as you mix. It sounds like this:

Audio example 28: Post-fader aux send

Audio example 29: Pre-fader aux send

Give it a try. Once you have your reverb device set up, send something to it by turning up the aux send. Now push the *pre* button off and on while moving your channel fader up and down. You'll immediately understand. You can even mute that channel entirely, and if the aux is set for pre-fader, you'll still hear that signal buried in reverb-land. A cool effect, but not a common one for most church services.

Now look at your reverb device and you'll see various controls and settings. Let's review the main ones you should be familiar with.

Digital Reverb Unit

Input level: Adjust this so you get a healthy signal level into the device. Watch the signal meter from time to time to make sure you're not overloading it. There's usually a green range (good) and near the top a red range (not so good).

Preset: Effects devices come loaded with different types of reverbs and other effects; each of these is called a *preset* or *patch*. Go through them and hear what each sounds like, then find one that seems to work for your situation. The most common categories are halls, rooms, and plates. A large hall patch will sound like a big auditorium, while a small hall sounds like, well, a smaller auditorium. Plates have a brighter sound; I tend to use these more often due to their crisp sound that doesn't get muddied in the mix. Experiment and keep it low at first, as people tend to overdo it.

Reverb time (RT60): Reverberation decays (fades away) over time as sound energy dissipates in the room. If you hit a snare drum with a long RT60 setting, it will ring out in the room for awhile, but if you shorten the RT60 the reverb fades quickly. Long RT60 times tend to become muddy and unintelligible, so keep these relatively short. Each patch you select already has an RT60 setting, but you should certainly fine-tune this to match your particular situation. Try 1 or 1.5 seconds—whatever it takes to keep it from building on itself as the song drives along.

Pre-delay: This is the time delay between the original sound and when the reverb is triggered. The delayed response is simulating the sound bouncing off the closest surface of a room, so longer pre-delays sound like they're in a larger room. With a very short pre-delay, reverb builds immediately, reducing the clarity of the sound. Keep these around 20-40ms, especially for your vocals.

Notice the difference between these next reverb examples. The hall sounds like it's in an auditorium, while the plate reverb sounds brighter. The RT60 decay time on the third example is far too long, resulting in a very muddy, dense mesh of noise that clouds everything. In this case, reduce the RT60 setting until it seems to fit with the song's tempo and overall sound. The fourth example has a shorter RT60 that works better.

Audio example 30: Hall reverb

Audio example 31: Plate reverb

Audio example 32: Reverb with long RT60

Audio example 33: Reverb with shorter RT60

Try this snare, dry first, then with a plate-like reverb added.

Audio example 34: Snare dry, then with a tight reverb patch

During a service, you can kill the reverb entirely by muting the return where the reverb signal is connected to the console. This will be either an aux return, stereo input, or an empty channel as we discussed earlier. Turn on the verb during music, mute it when they're talking. If you've got an instrument with way too much verb on it, just turn down the aux send on that particular channel. If the overall amount of reverb is too much, turn down either the stereo return (preferred), the input level on the reverb unit, or the master aux send.

You heard this one a moment ago, but it definitely needs the reverb aux turned down on the snare track.

Audio example 35: Too much reverb on the snare

Audio example 36: Entire mix with too much reverb

Audio example 37: Entire mix with a good balance of reverb

Audio example 38: No verb at all—can you hear the difference?

Notice that your effects device not only generates reverb, but lots of other cool sounds as well. Scroll through the patch list and you'll see various delays, flanging, chorus, panning, and just plain weird stuff. Some of this can be quite useful, such as adding a chorus or flanging effect to make a part more interesting.

Audio example 39: Dry guitar, then add flanging

Small amounts of digital delay on a track can really make it come to life. Here we take a basic rhythm guitar track and add a short delay. Use your headphones to really hear the effect.

Audio example 40: Dry guitar, then add delay (out, in, out, in)

Audio example 41: Same thing, but with the rhythm section

If you're running a mono mix in the room, short delays like this will cause phasing issues and thin out your sound; they work well when panned apart in stereo. Longer delays can be nice, though, so go ahead and experiment to hear what they do. We'll often run a delay processor and feed a few selective things to it, such as a lead vocal, percussion, maybe an acoustic guitar. Keep it pretty low so it doesn't call attention to itself. If you do any recording with your team, mix engineers will always use layers of delay and short reverb in their mixes, though in very subtle ways.

Uneven signal levels

When musicians play, their dynamic range usually varies quite a bit between soft and loud notes. Even during a relatively consistent rhythm part, such as a strumming guitar, fluctuations in a musician's playing can cause significant differences in levels. Same goes for vocalists, who may sing the verses normally, then kick into high gear on that dramatic final chorus. The result is that your instruments and vocals will jump in and out of the mix. Instead of frantically chasing faders, you can control and streamline this with a *compressor.*

A compressor is a signal processor that limits how much an audio signal varies from soft to loud. As the incoming signal gets higher, the compressor will reduce this so it sounds more even. Think of it as a volume cruise control for audio.

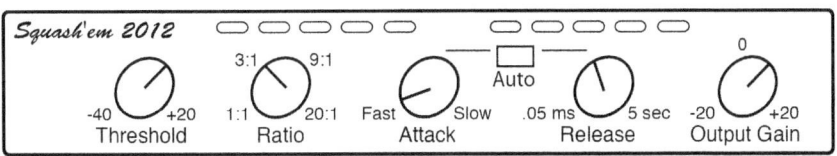

Here's a quick primer:

Threshold is the signal level at which the device begins to compress. Any signal that is higher than the threshold will get attenuated, meaning it won't be so loud coming out the other end. Threshold doesn't directly dictate how much compression will be applied, only the point at which the device starts doing its thing. A high threshold will only compress the highest, loudest levels in the signal, whereas a lower threshold affects most of the signal and is more noticeable.

Ratio sets *how much* the signal will be reduced. Once a signal goes over the threshold point, it will be attenuated at a rate set by the ratio. For example, if you set it at 3:1, every 3dB of signal beyond threshold will result in only 1dB coming out of the compressor (smaller dB numbers are quieter). Since these are ratios, multiples apply here, meaning if 9dB of signal goes over threshold, it will output 3dB.

A signal going over threshold triggers it to start compressing; how much compression actually occurs is set by the ratio. How *long* it takes to get to that compression amount is driven by the *attack* setting. If the goal is to attenuate 3dB above threshold, then a short attack will make this happen very fast. A longer attack setting means the compressor will take more time to get to that 3dB attenuation level. This is a big deal because it affects the tone of the sound as well—a really fast attack will lose the initial bite or attack of the sound, giving you a more rounded tone. Lengthening the attack time will allow that bite to come through before it begins limiting the overall dynamic range. The technical term for what you're shaping is called *transients*, which are the very first high frequency components of a sound that happens when the instrument is plucked or hit. So, turning the attack time up and down will give you a brighter or duller tone.

Once the incoming signal begins dying away and falls below threshold, the compressor will "let go" and allow the signal level to return to normal. How long does this take? That's the *release* time you set. A medium release is usually pretty smooth, whereas a long release may prevent the unit from recovering before the next musical passage begins. This obviously depends on the tempo and what they're playing. A very short release setting can cause the unit to cycle too quickly, resulting in a breathing or pumping sound. If your head is about to explode from all of this, relax a bit, turn on the "auto" switch, and move on. This lets the compressor figure out appropriate attack and release times on its own, so you don't have to worry about it.

After a signal goes through a compressor, it's usually got a lower overall signal level and might sound quieter in your mix. To compensate for this, all compressors have an amplifier at the very end where you can crank it back up a bit. Sounds contradictory. but what's happening is that once the compressor does its thing, you've got a more controlled signal with less dynamic level swings. Now you can take this streamlined signal and crank it up as needed. You set it in the mix and it won't jump in and out as much.

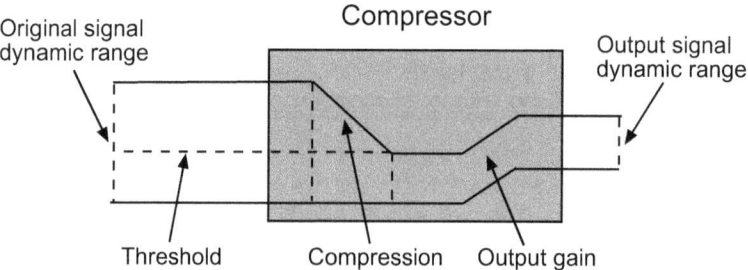

Listen and see if you can hear some of the differences the compressor makes. Even though I've exaggerated the settings, it will most likely take awhile to get it.

Audio example 42: Compression on saxophone (none, heavy, then medium)

Audio example 43: Compression on acoustic guitar (light, then heavy)

Audio example 44: Drumset with compressor on kick and snare (off, then on)

Probably the most noticeable result is the loss of the initial attack. The initial snare hit or guitar strum is the loudest part of the sound; if you have a fast attack setting on the compressor, it immediately holds this back. Listen how it takes a quick dip before leveling off a bit. Of course, different settings will change this somewhat as I described above, so put your headphones on and experiment by just turning knobs. Shoot for something that sounds like the medium settings in these examples; you don't want to overdo it because it'll kill the liveliness and dynamics of your mix.

You need one compressor channel for each single channel of audio you want to work on. If you want to compress the bass guitar, acoustic guitar, and the vocal team separately, you need a compressor for each part. You can, though, put a compressor on a group of parts if they're grouped on the console. So if the vocals are all routed to a sub-group, then on to the main mix, you can put a single compressor on that sub-group. To get started, try these settings and then experiment a bit for different instruments.

- *Ratio*: 3:1

- *Attack*: Keep it sorta fast, but not quite all the way

- *Release*: Medium

- *Threshold*: Slowly turn it up or down until you get just a few dB flashing.

- *Output gain*: Start at 0 (no change) and adjust as needed to provide a good level in your mix. You may not need this if you're only compressing a few dB.

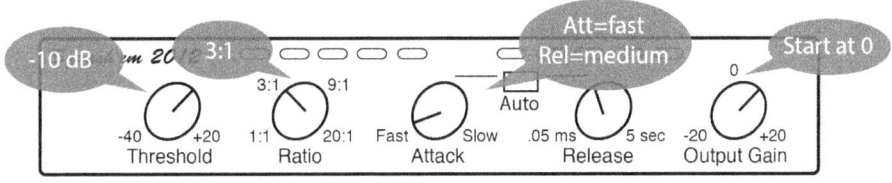

Usually we shoot for a few dB of gain reduction (GR). If the compressor meters light up like a Christmas tree, it's probably too much and will sound squashed, so raise the threshold. To see how to plug an external compressor in, jump over to *Level 4 Digging Deeper > Cable Connections*. Even if you're not the one to plug gear together, you should understand how it's all connected in case something goes wrong and you're the potential hero on the scene.

Suggestions and tips

Compressors tend to freak out novice engineers, sometimes to the point where they stay away from them forever. Jump in and experiment with the controls to see and hear what they do. Try different instruments and voice parts, even entire mixes. If you're fortunate enough to have a variety of compressors available, experiment to hear how each type flavors the sound. Similar to the concept of selecting the right mic, compressors come in many different models and affect the sound in unique ways. It's not all about controlling levels—light compression on your instrument and vocal parts sounds better much of the time, giving them a tighter, punchier sound. Overdo it, though, and it'll sound smashed, lifeless, and unmusical.

Of course, there's no rule that says you have to use a compressor at all. Use your ears and *listen*. Don't worry about what the numbers on the dials are showing. Does it sound wrong? Change it or get rid of it. Here are some ideas to help get you started, but it varies for different situations; hopefully you get the idea and can take it from there.

Kick

Compression on drums can help tighten the sound, fatten it up, enhance the attack, and make hits more uniform in the mix. Start with a 3:1 ratio and set the threshold until the GR (gain reduction) meter flashes 2-3dB. Set the attack a tad slower than the fastest setting to allow initial transients through (more pop), and set a medium-fast release time. The idea is to not crush it too much, losing the punch and depth you want from a kick, but instead to tighten it a bit. This will give you a punchier sound along with maintaining a fairly even level in the mix.

Snare

Use the same general concepts and settings as for a kick, but experiment with hitting it harder by lowering the threshold and increasing the ratio to 4:1 or so—you get a different tone that sounds really cool. Play with the attack time and hear how it enhances or diminishes the crack of the hits. Compression on a snare gives you that tight sound you hear on recordings, as opposed to the loose rattle you get standing in front of it.

> Audio example 45: Snare compression (out and in)

Entire drumset

If you've got mics on all your drums, you don't necessarily have to compress all of them. Focus on kick and snare, though on a digital board you've got compressors available on every channel if you want to experiment with the toms as well. Consider light compression on your two overheads mics since they're capturing everything; this can help control loud cymbals crashing all over the place.

With just two mics on the drums, you'll need a stereo compressor (one channel per mic). Set it for stereo mode (look for a switch in the middle somewhere), which means you set the controls on one channel and the other automatically matches them. Aim for just a few dB reduction on the meter.

Bass guitar

Start with a 3:1 or 4:1 ratio, then reduce the threshold until you see a few dB GR. Set the attack time to get the initial transient shaping you want (brighter vs duller), and the release needs to follow the length of their notes as much as possible (start medium, maybe medium-long). You want the compressor to pull back on the loud notes, such as when they start thumping, without making it sound constrained and thin.

Audio example 46: Bass compression (out and in)

Piano

This depends on the situation. For classical and jazz use very little compression—perhaps just enough to help even it out in the mix. Fairly high threshold (so only the really loud parts get compressed), perhaps 3:1 ratio. Look for the GR meters to flash just a bit. The attack should be a little slower than the fast setting with a medium release time.

Piano in a rock band functions a lot like a rhythm instrument, so compress it heavier to help place it in the mix. Run a higher ratio, perhaps 4:1, and lower the threshold until you see 3-5dB of GR.

If you're running two mics on the piano, use a stereo compressor as we described above for a drumset.

Audio example 47: Piano compression (out and in)

Vocals

Compression helps smooth a vocal part in the mix, especially for non-professionals whose performance will fluctuate quite a bit.

Start with 3:1 and lower the threshold just enough to flash a few dB GR when they're singing pretty strongly. Attack time should be fast. Keep the release time medium.

You can compress several vocal mics at once if they are assigned to a single sub-group bus. Connect the compressor by inserting it into the sub-group's insert point. However, if you've got one singer who has a higher dynamic range than the others (sings a lot louder), then that person will predominantly trigger the compressor, perhaps doing too much to the overall group. Put a separate compressor on their channel if you can, then use the bus compressor for the group.

One warning about over-compressing stuff, especially vocals. If you have a fair amount of gain reduction, which then calls for an increase in the compressor's output gain to get enough volume in the mix, what happens is that when the vocalist stops singing the compressor "releases" and allows the signal to return to unity gain. Guess what? This is exactly the same as cranking your channel faders up even higher. Enough of this and you'll get feedback. It'll only happen when they're not singing, as the compressor isn't reducing anything and the make-up gain control is still turning it up internally (a double-whammy). Lighten up on the compressor so it's not changing the overall signal level very much.

> Audio example 48: Vocal compression (out and in)

Guitars

Compression on electric guitars will keep them consistent in the mix. Depending on your church's style of worship, you might want the sound subdued just a bit so it doesn't call attention to itself. Try a 4:1 ratio, then set threshold until the GR meters light up a few dB. Depends on what they're playing, though. Chunking rhythm parts can be compressed more so they sit solidly in the mix. Solos and fills may sound squashed, so find a good compromise by raising the threshold.

Acoustic guitars can be treated about the same; it depends on what they're doing. If it's a rhythm part, more compression controls it in the mix. Lots of acoustic fills and leads can sound more musical with less compression. Use a higher threshold with a 3:1 or 4:1 ratio, and play with the attack time to shape the transients (brighter vs duller).

> Audio example 49: Guitar compression (out and in)

Main mix

Ok, you got me. This is not an instrument, but we often insert a compressor on the main mix. Compressing the mix will tighten the entire sound (a good thing); just don't overdo it or it will sound like it's being forced through a rubber hose. Keep a high threshold with a 2:1 ratio so the GR meters don't light up much at all.

Sometimes a *limiter* is installed between the console and the main amplifiers. A limiter is a compressor with a very high ratio. It's usually set at a high threshold as well; the idea is that high-level spikes or pops in the signal will trigger the limiter, which then clamps down pretty hard. This protects the system from damage, such as when your impatient guitar player keeps unplugging without warning.

Noise from open mics

Microphones should be off when not in immediate use. Sometimes it's easy to hit the mute button on each channel, especially if your console features a group mute function where you can kill several channels at once. Other times it'd be great to have something do this for you. Open mics with various noises and sounds coming through clutter the mix and may be noticeable to the congregation. If your pastor's mic is on and they're in the bathroom, it will certainly be noticeable to the congregation.

A *noise gate* is designed to "shut down" an audio channel when the incoming signal stops or gets really low. This is an over-simplification, but the idea is that a gate can keep background noises out until the person begins singing, playing, or talking.

A great example is guitar amplifiers. Ever notice what these babies do when waiting for liftoff? They hum and hiss—very annoying, and very easy to set a gate that will close down this noise when the player's not doing anything. As soon as the guitar begins playing the gate opens. Set the threshold just at the point where the noise goes away; any higher and you risk losing some of the quieter notes they happen to play.

Audio example 50: Guitar amplifier hiss with gate off, then on

Here's another situation. The snare mic has a pretty loud kick drum leaking through. After inserting a gate into the snare channel you hear much less of the kick. This cleans up your sounds and provides more control over each part.

Audio example 51: Snare w/ kick–gate off, then on

A noise gate looks very similar to a compressor; several controls are identical, though they trigger a different response.

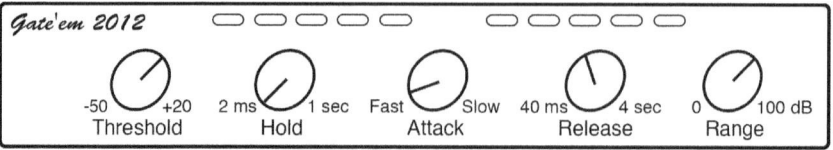

Threshold: determines the signal level where the gate will open; set it where it will open when the sound begins, then close again after it stops.

Hold: You'll probably never use this for worship services, so keep it as short as possible. For the curious, hold is a timer that closes the gate at a specified interval. A common trick was to set a short hold time, so when the timer ran out it would abruptly cut off the sound (gated reverb). That's enough for now...bad vibes from the 80's...

Attack & release: Once the signal rises above threshold, how fast does it open all the way so you hear the sound? That's the attack time, and usually you want it as fast as possible so it doesn't clip off the sound. When the musician stops and the signal drops below threshold, the release time controls how long it takes to close down the channel. Keep this somewhere in the middle and use your ears. If it's too fast, it'll cut off the sound too quickly. If it's too slow, you'll hear background noise after the main signal has stopped.

Range: I've used terms such as "close" and "open" to describe what a noise gate does, but it's not really an on/off switch. Gates attenuate signals, meaning they reduce signal level. For most applications, though, the end result is indeed essentially shutting down the channel as the gate is set to attenuate the signal very low. The amount of attenuation is set by the range control (also known as *floor*). If you set the range for only a few dB, you'll hear only a slight drop in the channel noise. Set it very low, say 50dB or more, and it will sound like it's completely off. Often we'll compromise and set it halfway down or more; it cleans up the track without some of the side-effects of a really low range where the gate can struggle to open and close in time.

Different engineers have their own preferences as to how they handle things. At my church we actually don't do much with noise gates on our musicians and vocal teams, preferring to use the group mute functions on the console. Gates can make your job easier during a service, but they require some experimenting to figure them out. If you have a digital console, you're in luck because they provide built-in processing such as gates and compression on each channel. Try it and see what you think. Doesn't matter too much how you get the job done as long as it facilitates a smooth worship experience for everyone. And by the way, a noise gate won't help prevent the pastor's bathroom visit from hitting the airwaves. Why? Because the gate will open as soon as there's sound coming through...maybe you should use that mute button pretty regularly.

Distractions

A sound system should be invisible. When all goes well the congregation will not notice anything but what's happening in the service. This comes to a screeching halt when a mic is turned on late, the singers are too loud, the pastor sounds like a droid from Star Wars, or feedback begins brewing. We try to prevent this stuff from happening, but when it does you need to know 1) what it is, 2) what might be causing it, and 3) how to fix it quickly. Here are the more common problems that you run into.

Feedback

Most people know what feedback sounds like, unfortunately because they've heard it so much over their lifetime. Sometimes it's obvious, other times it just hangs there behind everything else, annoying the congregation while the sound dude sits in the back blissfully unaware.

What is it? Feedback is caused by a continuous signal loop between a microphone, amplifiers, and speakers. When a microphone picks up sound from the speakers, this gets transmitted to the console and out to the main speakers...where it then gets picked up by the same mic again, and again, and again. The signal level builds each time, so it gets louder if left unchecked. Along with hurting people's ears, this can potentially damage equipment.

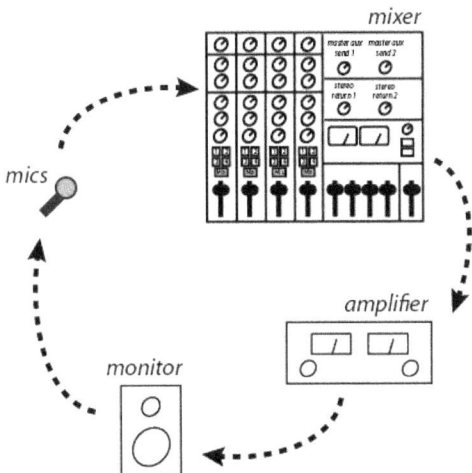

What causes it? Someone holding a microphone might be pointing it down toward a floor monitor or standing too close to a main speaker. The choir mics might be close underneath the main speakers. If you're using condenser mics, remember they have a high sensitivity for picking up sounds which lends itself to feedback issues. If you crank up the gain on a mic too much you'll hit the feedback threshold. I've also mentioned before that lots of open mics equals lots of feedback. Finally, all microphones should be behind the main speakers, but some installation situations don't follow this idea due to architecture, room layout, or just plain dumbness.

How to fix it? Move people and microphones away from the speakers and remind singers not to point their mics down into the monitors. Consider a dynamic mic if it will sound okay for that situation. Mute any mics that are not being used at that time.

When feedback happens during a service the first impulse is to pull the fader down, assuming you can find the channel that's having the problem. Know ahead of time which mics might be prone to this—you can usually rule out drum set mics and so forth, concentrating on the vocals, choir, or pastor mics. Then when it happens, you have only a few channels to examine. The signal meters on the console can indicate which channel is going into overdrive.

Sometimes a better solution is to adjust the EQ to attenuate the problem frequencies. If you're getting feedback or ringing in the same frequency range for different mics, then you've got a room acoustics problem that can be resolved by attenuating a narrow band of frequencies on your main mix. If only one mic seems particularly sensitive to feedback, try reducing the frequency area on the channel EQ. Careful, though, if you pull too much out you'll affect the tone of the signal and make it sound thin. The advantage of using EQ to solve a persistent feedback issue is that you can maintain the fader gain on that mic, rather than having to turn it down. One more solution is to install a feedback eliminator/suppressor device. These units plug into the main mix after the console and are sensitive to any frequency area that begins building in level above the rest of the sound. They will automatically attenuate that frequency band just like you did with the EQ. These are really nice to have, but the room problems ought to be dealt with first with a system EQ. I can't go into that here, so find a professional installer who can help get it done correctly.

Hum

If you turn on your sound system and hear a steady hum through the speakers, you've likely got either a ground loop problem or AC power leaking into your audio cables.

What is it? When you plug some of your equipment into one electrical circuit while plugging the other stuff into a separate circuit and they're connected with audio cables, odds are you'll get a ground loop. This happens a lot since churches power the console and other front-of-house equipment on one side of the room, while main amps, guitar amps, and keyboards get plugged into circuits on the other side.

What causes it? The reasons are fairly technical, but it has to do with power and how it is transmitted and grounded to earth. Two separate circuits will have different "potentials to ground", meaning a slight difference in how each circuit transmits electricity. The result is a 60Hz sine wave added into the audio signal, which sounds like a hum.

How to fix it? There are a couple of things you can try, but if they don't work you'll have to call an electrician or, better yet, a professional audio installer.

A common culprit is a keyboard or guitar plugged into a direct box. All direct boxes feature a ground-lift switch, which disconnects the grounding wire in the *audio* cable. This is safe, as the actual devices are still electrically grounded where they're plugged into the wall. Simply flip the switch to whichever position kills or reduces the hum.

Try to connect everything into the same circuit. This may not be very practical, however, because usually your front-of-house and backstage gear is a long way from each other. The one circuit also may not be able to handle the high load, especially with current-hungry devices like power amplifiers.

Do *not* break off the third pin on a power cord. It's there for a reason, and I'd rather you live long enough to enjoy the rest of this book and maybe go see the movie version in a couple of years.

Believe it or not, audio gear is not all created equal. Sometimes a particular device will inject a hum into your system and you can't do anything about it except throw it away. Disconnect everything, then gradually add each component until you hear the hum return—this will help identify the culprit.

Something easy you can check out is whether you have AC power lines running too closely to your audio cabling. Power runs at a much higher signal level than audio, especially microphones, so it "leaks" over into the audio lines. This is called *inductance.* Try to keep power cables away from audio as much as possible. If you must cross paths, set them perpendicular across each other. Run audio and power in separate conduit. I know it's nice and neat to run everything in one bundle, tied together with plastic cable ties, but resist the temptation. You'll regret it.

Years ago I was working at a Spyro Gyra concert where the audio and lighting guys could not get rid of a persistent hum. They finally went outside to the band's tour bus and connected a generator backstage to power the audio system. There's always a way…

> Audio example 52: Hum from a direct box

Buzz & other external noises

Along with power leakage, your system has the potential to draw in lots of different noises from outside sources.

What is it? Analog audio cables are subject to interference from all sorts of places. Room lighting on the same circuit can wreak havoc with this, as well as HVAC systems, refrigerators, and general RF (radio frequency). For years my church featured some great gospel radio and TV programs on Sunday morning…only they weren't actually supposed to be part of the service. It was all RF leaking into our audio lines. Some of you have been there when the pastor's mic suddenly got hijacked by truckers driving by talking on their CB radios. Still sends shivers down the spine.

What causes it? Lots of things. If other building appliances, such as HVAC and refrigerators, are on the same power phase or circuit you'll have a buzz problem. Lighting dimmers can be nasty, too. If your audio cables are not properly connected you invite outside RF from radios, cell phones, and so on. Lastly, even some of your audio gear can be the problem due to poor design—a good reason to buy quality equipment.

How to fix it? An electrician can help isolate power for non-audio gear in the building. You'd like to have totally separate, audio-only power circuits if possible. Digital snakes that use Ethernet cables are a huge improvement over traditional analog cabling as they are immune to such noise issues, so this can fix the connection between your stage mic inputs and the console. You still have to worry with instrument DIs, mic lines, etc.

You or your installer should check all your cables and connections and see what you can find; maybe an audio ground has broken loose (the shield in the cable which is connected to pin 1). Inside the connector, the wires should be twisted as close as possible up to the connector pins. Sometimes the culprit is in the equipment itself, making it very difficult to resolve. The story goes that the real reason Adam bit into that apple was that he was completely frustrated trying to solve an RF problem in their home theater.

Vocal p-pops

You remember sitting in a lecture hall or sanctuary listening to someone speak, when suddenly a gigantic "BOOM" thunders through the system?

What is it? Pops are generated by hard consonants, mainly "p" and "b", forcing a huge amount of air pressure into the microphone. You'll hear a pop or boom, sometimes loud, other times more subtle. These are mostly low-frequency sounds.

What causes it? First check to see if anybody's looking at you. If the coast is clear, put your hand in front of your mouth and start talking, using lots of p's and b's. You'll feel puffs of air at each of these consonants, focused downward from the mouth. This blows across the microphone's diaphragm (the part that captures incoming sound waves) and overdrives it.

How to fix it? Proper mic technique can help a great deal. Have singers position their mics down just below mouth level, facing upward against the front of their chin—the air will mostly go above the mic. Don't let them go rock star on you, holding the mic straight out from their mouth. This is the worst position for pops. Placing a mic off to the side works also, which is one reason you see a trend toward headworn mics for pastors and singers. Many mics, and all console channels, have a low-cut filter switch. These attenuate low frequencies, usually around 75-80Hz and below. This will reduce most, but not all, of these pops, as their frequency content extends too far up the scale. I have filters turned on for all my channels at church except for bass and kick drum. Not only does it reduce pops, but it eliminates lots of low frequency rumble and garbage. Listen to the next example and hear how turning on the filter not only reduces the pop, but also cleans up the bottom end considerably.

> Audio example 53: Using a low-cut filter to eliminate vocal pops

Late mic cues

This one is pretty simple. Make sure the mic is on just before the person speaks or sings, not after they've already started. This is a huge distraction, much worse than guitar amp noise. Now, when they're done, mute the channel before they set the mic down or put it back on the mic stand so you don't get a thump through the system.

How to fix it? Pay attention.

Overall volume

Yes, volume can be a distraction. If the music or pastor is too loud, it's uncomfortable for the people listening. Obviously, if it's too soft people can't hear. The challenge, though, is that some people will think it's too loud while others can't hear a thing, and most of them will be happy to let you know. Find that happy medium while remembering that a sound system should not call attention to itself.

Creating mixes for other purposes

Along with the main sanctuary mix, you may need to provide mixes for hearing assistance systems, distribution throughout your church facility, web streaming, recording, and so on. You don't want to send a copy of your main mix because it won't sound the same for these other destinations. The mix you craft in the sanctuary takes into account the sources you hear acoustically in the room, your microphone sources, and the room's acoustics. Step outside and you don't get this.

Our church has these cheap ceiling speakers located out in the main hallway and in the nursery. And after many years of "I really need to do something about this" they still sound pure awful. Remember the graphic of our church mix I discussed earlier, showing the difference in fader levels between acoustic and electronic sources? This really comes into play here, because the mix we're sending throughout the building has lots of vocals, keyboard, and bass, and almost no piano and drums. It's also missing the natural reverberation in the sanctuary that provides life and ambiance. The naked cheese-ball mix we're sending out has none of this, making it even worse. Maybe I'll get it fixed before you read this.

You must set up separate mixes for these situations. It's not hard; remember how to set up a monitor mix for your worship team? We use aux sends on the channels. Think of it this way—aux sends are really independent mixers, grabbing a copy of whatever comes through the channel and sending it to a master aux send. Here all your channel aux sends are combined into a single mix and then routed out of the console to their destinations. Since your console has multiple aux sends, we use one send per mix. For example, we can use aux send 1 on all channels for the stage monitors, aux 2 for a hearing assistance system, aux 3 for a building feed, and maybe aux 4 for recording the service. The following diagram shows the signal flow for a stage monitor mix using aux send 1.

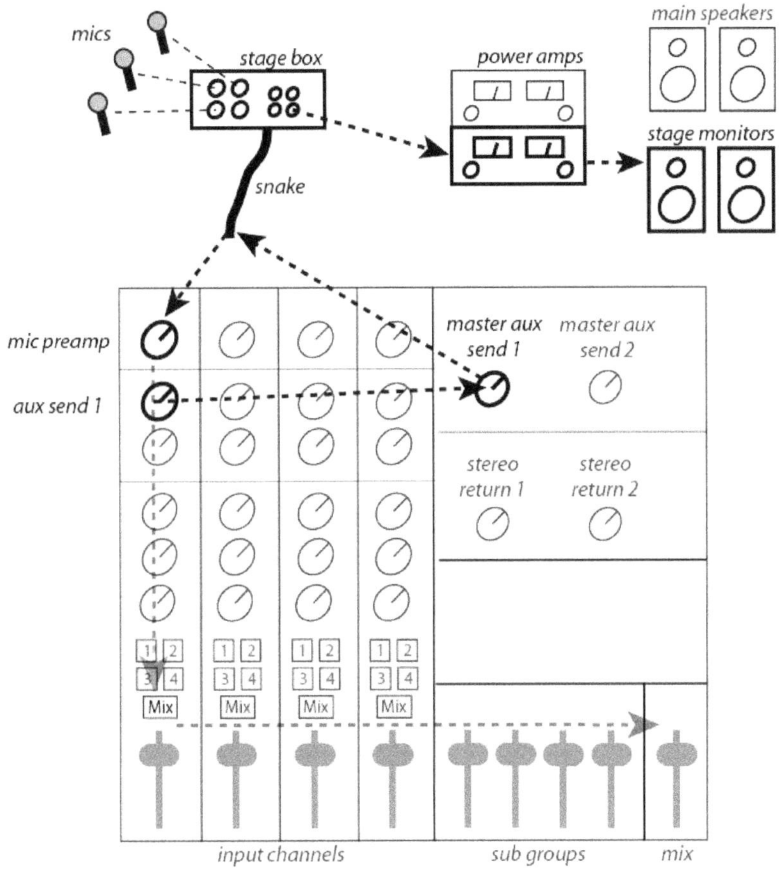

mics

stage box

power amps

main speakers

stage monitors

snake

mic preamp

aux send 1

master aux
send 1

master aux
send 2

stereo
return 1

stereo
return 2

Mix Mix Mix Mix

input channels

sub groups

mix

The only difference for setting up a hearing assistance mix is to route the output of the console aux send to your HA transmitter. Here's the signal flow for a hearing assistance system:

Channel aux sends ➡ Master aux send ➡ HA Transmitter ➡ Receivers

The first catch is to make sure the auxes you're using can be set *pre-fader*. Aux sends can grab a copy of a channel's signal either before the channel fader or after. If an aux feed comes after the signal goes through the fader, then every time you adjust the fader level for the main mix it also changes the aux send level to your

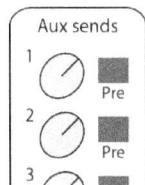

other mixes. Not good, but if you get your aux feed *before* the fader, then you can mix away and not affect these other destinations.

How do you get these pre-fader? Some auxes are permanently set to "pre" by the console manufacturer. Other auxes will have a small button labeled "pre", which when selected will provide a pre-fader send. The short rule of thumb is that if you're creating individual mixes separate from the main sanctuary mix, always, always, always use pre-fader. And remember, any changes to the channel mic preamps will affect *everything* else, including these aux mixes, no matter whether it's pre or not.

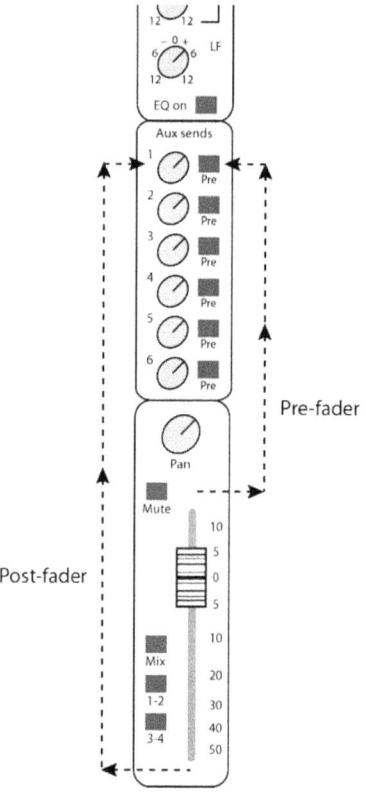

Even though it looks like an aux send gets signal at the aux send section, it actually grabs a copy from the fader section, either before the fader or after. It then sends it to the corresponding master aux send.

The second catch is that you have to listen to each of these mixes to make sure they're balanced. You really can't assume that since you checked the hallway mix a couple months ago everything is still good. You can listen to master aux sends using headphones and the solo buttons found next to the master aux control. Better yet, go out into the hallway or nursery and listen to what it sounds like. Grab one of the hearing assistance receivers, plug in some earphones, and hear for yourself what those people are getting. You'll probably be shocked and dismayed and wonder why they haven't slashed your tires before now. It's easy to forget what others have to endure when you're not in their situation, so go find out.

Recording your services (and other things)

Chances are your church records services for at least one of a few possible purposes. Sermons or even entire services can be burned to CD or downloaded from a website, allowing shut-ins or folks who are out of town to listen from home or on the road. Worship teams can produce CDs of their own music for the congregation to enjoy. There are some legal issues to consider, but I'll put that at the end so you don't fall asleep too quickly.

Recording sermons

This is easily accomplished by using an auxiliary send that is routed to your recorder. We just discussed using aux sends for independent mixes, such as for monitor feeds and hearing assistance systems; the same principles apply here. If you take a copy of your main mix and send it to your recorder, any adjustments you make for the sanctuary mix will change your recording levels. This won't work, so use a separate aux send on the pastor's mic channel and set it to pre-fader. Run this through a compressor to flatten the dynamic range; when people are listening in their car or jogging through the park, they can't hear what's being said if the signal volume is jumping up and down. Compress it fairly heavily for a consistent, smooth level; try starting at -10dB threshold and around 4 or 5:1 ratio. Your goal is to see several dB of reduction on the meter during louder passages, so turn down the threshold as needed. Lastly, turn on the low-cut filter on the mic channel. This will eliminate most low-frequency rumbling and noise, giving you a clean voice sound.

Here's the setup:

1. Plug the pastor's mic into a channel on the console.

2. Turn on the low-cut filter on that channel.

3. Select an aux send and set it to pre-fader.

4. Turn up the corresponding master aux send.

5. Connect the master aux send output into a compressor.

6. Connect the compressor output to your recording system.

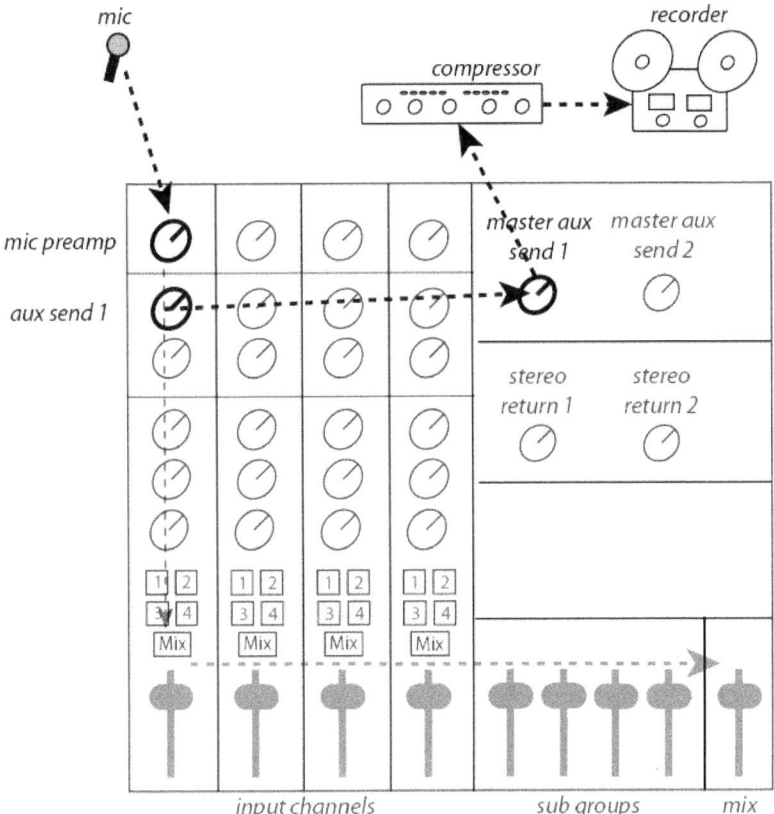

Watch your signal level throughout the sermon; it should be lighting up the meters pretty solid. If it keeps bumping the top of the recorder meter, pull it down using the channel aux send or your recorder input control. Of course, if it's too low on the meter, crank either of these up. Is it jumping all over the place? Increase the compression by lowering the threshold or increasing the ratio.

Make sure you burn a CD or upload a file and listen to what you've got. Sound too squashed? Raise the threshold and back down on the ratio a bit on the compressor. Do you have to crank the volume to hear anything? Your recording levels are too low. Does it sound distorted and fuzzy? You're overdriving something, so lower the aux send or the input level on the recorder. It'll take some experimenting to get it pretty close, and of course this will change a bit with different people speaking. Also keep in mind that any changes to the channel mic preamp will *always* affect your sanctuary mix as well as any aux send mixes.

Audio example 54: Distorted sound

Take a look at the following waveform diagrams to see the difference between a signal recorded too low, too high, and how a compressed signal helps you out.

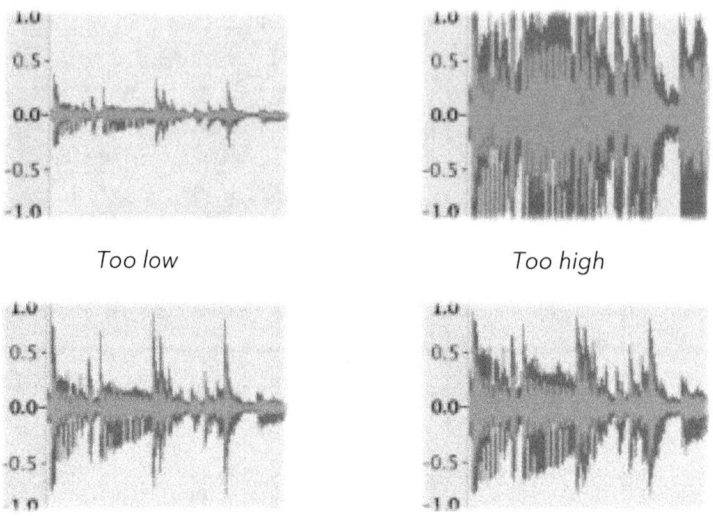

Too low   Too high

Good level with wide dynamics   Good level with compressed dynamics

Here's a spoken word recording with very heavy compression—you see almost no variation in dynamic range at all. You won't have to worry with the volume going up and down, but the ear can become fatigued listening to this for 30 minutes. Suggestion: back off the compressor a tad with a higher threshold and lower ratio, and always use your ears to make a final judgment.

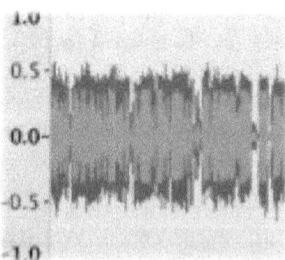

Should you EQ the mic? Depends. With this setup you are using the same channel EQ for both the sanctuary mix as well as the recording. The main service mix is arguably more important, so you probably want to fine-tune the mic EQ accordingly. Hopefully, if you don't need drastic measures to compensate for bad acoustics, mic choice, or mic placement, your EQ settings should be minimal. If you find it necessary to tweak something on the recording and you're using software, you can easily do this after the service.

One more thing you can do to improve your recordings is to make sure the signal is as high as possible without distorting. During the service, we almost always keep signals a bit lower than maximum so we don't accidentally distort the recording. Once you're done, however, and if you are recording into software, you should *normalize* the file. Select the entire waveform and apply the normalize function (it'll be somewhere in your menus). The software will scan the entire recording to find the highest level peak, then subtract that level from maximum. In other words, say your recording peaked out at -3 along the way. "0" is the maximum level in a digital system, so you've got 3dB of difference (what we call *headroom*). The software will then increase the entire file level 3dB. This does not change the relative levels from soft to loud (that's for the compressor to take care of). But, if you've got some leeway before maximum, it will raise everything equally. The result is a louder recording, which is a good thing.

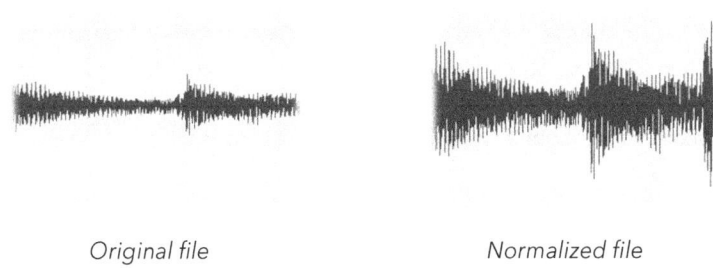

Original file   Normalized file

Recording the entire service

There's not a lot of difference between recording sermons and the entire service. You still want an independent mix using an aux send, though if you want a stereo recording you'll need two aux sends (or a stereo send if your console has them). Turn up the aux send on any channels you want to record, including instruments, vocals, whatever. Again, use pre-fader so any channel fader changes for the sanctuary mix won't affect the recording. Listening to this mix on headphones is crucial because there's a lot to balance. This will be different from your sanctuary mix, so you cannot assume that everything's fine. EQ becomes more of an issue, but again your main mix is first priority, so you just hope it doesn't wreck your recorded sounds. Setting the compressor is also a bit more involved because music triggers it differently than a single voice. That heavy compression for the sermon will be too much for music, so you need to find a compromise.

Recording a service on-the-fly can be challenging. Instead of chasing down a decent mix as things move along, consider capturing all your mic signals and parts individually. Then you can create a much better mix on your own time. This involves multitrack recording software and hardware, which I'll describe a bit later.

Recording your worship team

If your worship team wants to record an album, don't do this during a service unless you really want that "live" feel. You need to control the environment, avoiding the various noises, mistakes, acoustic issues, and other bad stuff that always happens in a live event. You can experiment with microphone choice and placement. Various parts can be recorded at different times, just like we do in the recording studio. You can record multiple takes as the musicians correct mistakes and work on intonation. This is not the place for a detailed discussion on recording sessions (you could read another book of mine…sales plug…just sayin'), but take what we've covered throughout this book and fine-tune it. Use your ears and carefully listen to that mic choice, placement, guitar tuning, and so on. Pay more attention to the acoustic environment, experiment with stage locations, space acoustic instruments farther apart to reduce leakage. Capture this with a multitrack recording system so you can tweak each part and bring it all together. This type of recording is lots of fun to do, but it also requires a tremendous amount of time to do it well.

Multitrack recording

Multitrack recording is a lot more involved than running aux mixes straight off the console. You need a system that can capture multiple channels of audio, record it, process it, play it back for mixing, and record a final stereo mix of everything. These days we most often use multitrack software on a computer, but this requires certain hardware to convert mic signals into something computers can handle. Here are a couple setups to give you an idea of what's going on.

Using the console

In this setup, connect your mics into the console as usual. You can EQ on the channels and insert processing such as compression (though I would do this later when you're mixing, not recording). But instead of using the main mix like you do during a service, you've got to take each of these separate channel signals and feed individual tracks of an external recording system.

One way is to assign mics to your sub-group buses and connect each bus output to individual tracks of your recording system, but many live consoles do not have enough buses to match the number of mics you're running. Let's say your console has 4 sub-groups. If you are recording up to four mics at a time, this works fine. Plug a mic in, route it to bus 3, then connect that to track 3 of your recorder. But that's pretty limited since we typically lay down several parts at once, such as an entire rhythm section. This involves separate mic inputs for bass, guitar(s), keyboards (stereo), and drums (several mics). So, we need a way to feed several individual mic channels to the recorder.

Many consoles have a *direct out* jack for each channel. These are found on the back of the board; connect a cable from this jack to the input of your recording system. This requires a recording system that can handle several inputs at once—more on that in just a second.

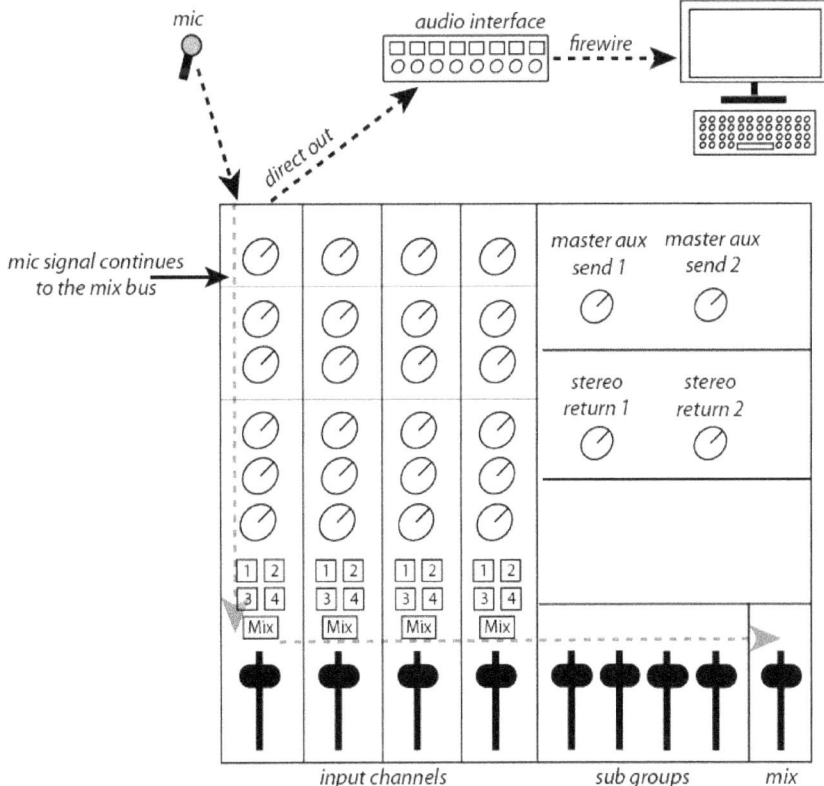

Another way for sending mic signals directly to a recording system uses the *insert jacks* on each channel. Yes, we normally use inserts for sending a signal to a processor, then returning back to the channel. But, if you half-jack a cable in an insert, it grabs a copy of the mic signal but allows the original signal to keep going through the console. Huh? Slowly plug an insert cable (or any balanced 1/4" cable) into an insert jack and you'll notice there are two clicks. Now plug it in until you hear/feel the first click, then stop. You've now tapped a copy of the signal that can be connected to your recorder, while the main mic signal continues through the channel and is routed to the mix bus as usual. Connect the send or tip of the insert cable to your recorder; don't do anything with the return—just leave it hanging. Both methods work the same, but if you have direct outs, these are best so you can save your inserts for connecting processors.

Using external audio interface mic inputs

The most direct and cleanest way to make recordings skips the console entirely. Plug your microphones into a multichannel audio interface that has several mic inputs; this device then outputs a single data signal (USB, Ethernet, etc) to a computer running multitrack recording software. All of the processing such as EQ, compression, reverb, and fader balancing is handled in the software. To hear what's going on, take the monitor outputs (stereo) from the audio interface and connect to a set of quality studio monitors or headphones. You can also feed this into your main console if you want to hear it over your main speakers...just don't do the final mix with them!

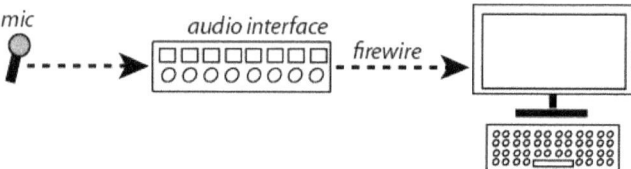

Audio interfaces

These devices take analog signals, mic and line, and convert them to digital for transmission via Ethernet or USB to a computer system. There are tons of audio interfaces to choose from, so where do you start? If you're only making a stereo recording via an aux send mix, a simple two-channel interface works fine. For multitrack recording you've got to have a mic preamp input for each microphone. This usually means an interface with at least eight mic preamps, and often folks will buy two or three of these units to increase the number of tracks that can be recorded at the same time. Be careful to make sure the unit you're about to purchase has what you need—many interfaces are marketed as featuring "eight inputs", but really only have perhaps four mic preamps and four 1/4" line inputs. If you're plugging a microphone into the box, it's got to have a mic preamp. If you're interested in getting the best sounds possible, don't go cheap on the interface. These can be fairly expensive, but a quality interface makes a huge difference.

Your best option is to visit the local music store that sells this stuff. Hopefully you can find someone who knows what they're doing, and they can show you various options and how it works.

Digital consoles connected to computers

Digital consoles can connect directly to a computer via Ethernet or USB. They usually offer a software application that captures all individual channel signals coming into your console and records them as a multitrack file. This is great for two purposes: making a recording that you can mix later, but also for recording rehearsals or performances for what we call "virtual soundcheck". If you record your worship team playing through a few songs, then you and your audio team can come back later to practice mixing and work on settings without the band having to be there. Just play the recording back through the console channels as if they were live. For many of you, this is the way to go. You've already got the hardware and software, so free is good, right?

Legal issues for recording

And now the fun stuff. Over the years I've asked various churches whether I can walk into a grocery store, grab a few cans of beans for the local food bank, and walk out without paying. "Of course not", they gasp. "That's stealing." "But it's for God!" Nope, doesn't matter. It's illegal. So why do you print copies of music and make CDs to sell at the annual fall festival? But isn't that different? Of course not, and the church needs to set an example of doing things the right way. Why am I telling you all of this? Because you'll be doing the recording. Other church members won't have any reason to stumble across anything as mundane as copyright law, so you're the Chosen One. So, share this with your team and church administrators so everybody knows what can and cannot be done.

Churches need to know what can be legally performed, recorded, and distributed as well as what it takes to do it right. Much of what is presented during a service is owned by somebody who deserves to be compensated for creating it. Music, videos, film clips, graphics, artwork, and other items are protected under United States copyright law; using such works in various ways requires permission, known as a *license*, that involves restrictions and financial obligations. Here's a very brief rundown of what you can and can't do.

Music

Unless you wrote the song, somebody else owns it, which means you must pay for the right to perform it. The only exception is for really old songs for which copyright protection has expired; they are considered to be in the *public domain* and can be used freely. There are two main categories for music licensing for church services:

Performance

Singing music in a service is termed a "public performance". This often includes providing lyrics on an overhead screen, making print copies for the worship team, and perhaps including them in the bulletin. Sounds like a lot to deal with, but thankfully it's a one-stop solution through Christian Copyright Licensing International (CCLI). For a really low annual fee, you have the right to sing and copy anything in their catalog. This includes most any hymn and worship song ever written; it does *not* include arrangements such as choral and orchestral works, nor does it include secular songs.

Recording

You cannot make and distribute a recording of someone's music without paying a license fee. Although this is a separate license under copyright law, the standard CCLI license allows you to record live music (you cannot use commercial recordings, meaning the ones you buy on CDs or as downloads) as performed during your service. You can distribute these freely (or for a small fee) to shut-ins, missionaries, and anyone who's part of your church. This does not include podcasting or streaming from the church's website; this requires a separate license (see below).

Want to make CDs or downloads and sell to the general public? You can't without the proper license, and CCLI does not cover this. I won't go into detail here; this type of activity is termed *mechanicals* and is handled by the Harry Fox Agency (harryfox.com). Contact them for information.

Other media

Showing videos and film clips is also known as a public performance under copyright law. CCLI offers a separate video license for presenting these during a service.

Podcasting and streaming

CCLI also provides a license for podcasting and web streaming of live services. Again, you cannot include commercial recordings, just the live performances.

Sharing commercial recordings to learn new songs

It's really handy to share the original commercial recordings of songs for your team to learn before rehearsal. For instance, I used to buy songs from iTunes or import them from CDs, then post them on my private church worship website so the group could listen on their own time. Nobody cares if you want to play these during the rehearsal, but making copies or posting them on a public website is not allowed. CCLI has a church rehearsal license for this purpose. Again, really cheap and worth it to do it legally.

Sermons

Unless your pastor is stealing someone else's sermon material, you don't have to pay a license to record, stream, or make copies of sermons. What a deal. So if you don't include music or other media in your recordings, no licenses are needed.

Is your head spinning yet? I've taught copyright law for years and yes, it gets dreadfully arcane and technical. The bottom line is that CCLI can solve most of your needs, so check out their website for more information (ccli.com).

Here's a summary of what you can legally do with the proper CCLI license.

Activity	License needed
Perform live music	Standard church copyright license
Make print/electronic copies	Standard church copyright license
Record live service music	Standard church copyright license
Distribute service recordings	Standard church copyright license
Stream/podcast services	Streaming/Podcasting license
Show video/film clips	Church video license
Distribute commercial recordings for rehearsal prep	Church rehearsal license

Solving problems

Ever feel like throwing the console across the room? At somebody? You're not alone, and there's probably a support group somewhere out there for all of us. Much of what you're running into, though, is actually fairly commonplace and easy to diagnose. Just take your time, don't freak out, and step through the possibilities. This list will not only get you started, but will also help you look like a genius next time things run amuck.

Sound and mix issues

I can't get any sound

Happens to all of us. Sometimes it's something really simple and stupid, other times something's broken. Most of the time it's user error. The more you know how the system works, how it's all connected, and understand signal flow, the quicker you can diagnose what's wrong and fix it.

First, start with the non-techie items:

- Is anybody actually using the mic? Duh...have them speak or play.
- Sometimes they're not singing or playing loud enough to get a decent level, especially with dynamic mics like the Shure SM58.

Check your channel settings on the mixer. Consoles have signal indicators or meters on each channel, so look for this to flash. If not, start here:

- Phantom power on (for condenser mics and active direct boxes)
- Mic preamplifier turned up
- Channel fader up
- Channel on (or mute switch off)
- Assigned to the mix bus (or a subgroup)
- If routing to a subgroup, group fader up, not muted, assigned to the mix bus

If you're getting signal to the channel, as indicated by a flashing signal indicator or meter reading, and still not getting any sound at all, move over to the master section of the console.

- Main mix bus muted? Shouldn't be.

- Main mix fader up to unity or 0.
- Are there processors patched into the mix bus, like an EQ or compressor? If so:
    o Is it turned on?
    o Is it patched correctly to the console?
    o Are there any gain/level controls turned down?
    o Try disconnecting it completely; if you get sound it's a processor setting or cable connection.
- Are your amps turned on?

It can also be a microphone cable or connection on stage.

- First, remember to mute the channel on the console before plugging, unplugging, and moving microphones.
- Is the mic plugged into the correct mic panel jack?
- Is the cable connected fully and securely to the mic and stage panel? XLR connectors (the 3-pin type) will click into place.
- Is the mic's attenuation pad switched on? This will drastically reduce signal level, so turn it off.
- Try replacing the mic cable.
- Try a different channel # on the snake and console.
- Try a different microphone.

And lastly, it might be something bigger that's broken, such as your console. A few years ago I was visiting my in-laws' church and was asked to help figure out why they couldn't get sound. One look at their mixer showed that it had no power—no lights, LEDs, nothing. I pulled the fuse out (located adjacent to the power cord) and discovered it was blown. Easy fix in this case, but not easy for the church's sound guy to know where to look.

Man, that guitar amp is really loud and I can't get it out of the mix.

Yeah, they do that. This is where your charm and personal relationship comes into play as you find tactful ways to have them turn it down, face it away, or even get rid of it and use the monitors. If there's a way to have them sit in the sanctuary while someone else plays, they can hear for themselves how it impacts the overall mix.

Having said this, guitar amps usually need to be turned up pretty high to get the desired tone. So, try these ideas:

- Turn the amp facing away from the congregation, maybe even directly toward the guitar player so they can hear it easily. At least turn it sideways away from people.

- Put the amp in a back room or closet. This reduces the amount of sound leaking out onto the stage. Place your mic in there with the amp, and don't forget to add guitar into the monitors since they can't hear it acoustically anymore.

- The best solution: Get rid of the amp entirely. Try an amp simulator that features sounds modeled from lots of different actual amplifiers. Dial in that Marshall stack sound without having to drag the beast across the floor.

That guitar amp has a lot of hum and noise.

The basic noise factor with a guitar amp setup is the amp itself. When you amplify a signal it generates noise. Guitar amps have to crank up the signal a *lot*, so you get loads of noise. Try putting the amp in a back room or get rid of it entirely and use an amp simulator.

Electric guitars use un-balanced cables to plug into their amps. These setups are prone to RF (radio frequency) interference, meaning other noises get inducted into the audio signal through the air. Some of this, strangely enough, is directional, meaning that the musician can turn in different directions and often change the character and level of the noise. You'll have to experiment to find a decent solution. Shorter, high quality cables are better, so that's an easy thing to try.

A noise gate is a great solution for eliminating amp noise while the musician isn't playing. That's when it's most noticeable, so even though the gate has to open while they're playing, at least the music will cover most if not all the noise. Insert the gate directly onto the guitar channel and set the threshold just above where the noise goes away. Make sure the attack time is as fast as possible so it opens as soon as they strum a note.

The band doesn't seem tight–it doesn't feel solid.

Either they don't know the song, they aren't used to playing together, or they can't hear each other really well. The last item is trickier than you might think. For example, at my church, a lot of what the band hears in the room are instruments and vocals bouncing off the opposite wall of the sanctuary. This, of course, is delayed in time from when the notes are actually played, so the worship team can't possibly play "in the pocket" together. To avoid this problem, our band members wear headphones and use personal monitor systems (earphones work well as long as they block out external sounds). This allows all of us to hear everything "at the source" with no delays. Trust me, it makes a big difference.

The singers can't seem to stay in tune.

There are several reasons for this, the simplest being that they just can't sing. Solution—your worship leader should get rid of them (nicely). Otherwise it's usually a monitoring/hearing issue that stems from the group of singers working together or what they are hearing from the monitors. Some singers work better together than with others; the "wrong" combination of vocalists can be a real issue. Some singers get thrown off key by someone they're standing beside; a weak or overly dominating vocalist can cause havoc in the group. The other issue might be getting what they need into the monitors. Singers need dominant, pitched instruments for reference, so piano and acoustic guitar work well to crank up in the monitors. No matter how exciting that cow bell part is, maybe they don't need quite so much of it in order to sing. While they're running through a song, go up there and listen where they are standing so you hear exactly what they're experiencing. You might be surprised, and you certainly will be more sympathetic.

The singers don't sound as full and loud as they usually do. Should I EQ something?

Is it the first run-through at 8am? Well, duh. Are they learning a new song? Vocalists and musicians never play with their normal gusto when they're unsure or just warming up. Don't make any changes based on this until you get a feel for how the entire rehearsal set is going, then decide if you need to nudge the faders up a bit. Assuming you've previously spent a good amount of time setting EQ for each vocalist, I wouldn't do anything drastic. Sometimes things sound "off" for whatever reason, so maybe try some small tweaks to help that morning. Remember to revert back to your regular settings afterward.

Keyboard is loud one minute, then you can't hear it at all.

Keyboards (synthesizers) provide lots of different instruments and sounds to choose from, but the volume level can change dramatically as the musician changes patches. On top of this is the fact that some sounds are meant to be front and center, others only for background fill. This requires coordination between you and the player.

Either the keyboardist offers to adjust volume levels as they change sounds, or you must keep track of where you need to turn the fader up and down. Of course, the musician can't tell where it needs to be set for the main mix, so work together to come up with a plan.

I keep getting feedback.

One of the reasons audio engineers can't wait to get to Heaven is that it's the Land of Perfect Sound; at least that's what the travel brochures claim. Until that rapturous day arrives you can arm yourself with some possible solutions. First, you need to know what causes it—feedback is a signal loop from microphone to console to amplifiers/speakers to microphone and on and on. When a mic picks up a sound from the speakers or monitors, it sends it back to the console. If that signal is then routed to the same speakers or monitors, then the mic picks it up again and the cycle builds until you hear the howl. If left unchecked, it will annoy everybody in the room and eventually damage something. Try the following:

- Move mics closer to their source so you don't have to turn up the mic preamp as much. This is a much bigger deal than you might think. Sound attenuates (drops) roughly 6dB every time the distance doubles. Think about that—compare a mic two inches away from your mouth with the mic being held a foot lower. From two to four inches, from four to eight inches, then another four or so, and the signal you're getting to the console has dropped over 13dB. Look at your channel fader markings and see how much you have to push it up to compensate. That's why you're getting more feedback. Remind your singers to hold the mics close, consistently, and you can lower your overall gain.

- Make sure the singers aren't pointing the mics toward the monitors or speakers. Monitors should be located in front of the singers, which would be behind the microphone pickup zones (remember *polar patterns?*). Cupping a hand over the microphone is a common reaction, but this also increases feedback as it captures and focuses sound from the monitor into the front of the mic (just like your DirectTV dish).

- Hopefully your main loudspeaker system is installed slightly in front of your stage area, so that all mics are pointing away from them. This doesn't always happen, and it's difficult to change such a configuration in the room.

- Make sure you are using *uni-directional* or *cardioid* microphones. Most microphones fit this description, and all it means is that the mic will pick up sounds from the front of the grill while rejecting most sounds from the sides and rear. So, when you are holding a mic toward you, the back of the mic is facing the monitor on the floor and will not pick up that monitor's sound as easily.

- The most common instinct for sound operators is to lower the fader level on the mic channels. This will work, but if you know something about EQ and frequencies, you can use the channel or system EQ to reduce problem feedback frequencies in your particular room. This requires determining what frequencies are feeding back and reducing those on the EQ. If you find that many of your mics tend to ring at the same frequencies, use a system EQ to reduce them. A *system EQ* is simply an EQ that's used on the main mix; it could be an outboard EQ processor that's connected to the console's mix bus or it could be built right into the console. Audio engineers, when setting up a sound system, will ring out the room by raising mic levels until they figure out which frequencies are problematic in that room. They then use the EQ to attenuate (reduce) those points and get a smoother response.

- If it's the vocal monitors that are feeding back, insert an EQ into the monitor signal. If you're running an aux send for your monitor mix, take the master aux send output from the console and plug it into the EQ. The output of the EQ then goes into your monitor return in the snake to feed your system on stage. Digital consoles give you EQs on each bus (aux). Use this EQ to pull down frequencies that seem to ring out more often.

- Remember that condenser microphones are more sensitive, and therefore more prone to feedback.

- Lots of open mics = lots of feedback. Reduce the number of mics you're using at any one time as much as possible.

How do we ring out a system to avoid feedback?

The simple version: turn on your mics and slowly raise the faders until you hear a squeal or howl begin to build. Now, try to hold that without it getting worse by adjusting your faders a bit. On the EQ that's patched into the main mix (or monitor mix), start pulling down certain ranges until you hear it diminish. With some practice you can learn where these frequencies are. Don't cut too much or you'll wreck the overall sound; your installer should have taken care of this, but maybe it's time for them to do a maintenance checkup.

I hear a ringing sound... I turn down the high frequency EQ, right?

One Sunday one of our leaders was speaking from the pulpit and we had this persistent ring. The common impulse is to diagnose this as standard feedback, curable by pulling down the fader or reducing the high frequency EQ. However, in this case the ring was around 500Hz—way down past the middle of the frequency range. No high frequency control will touch this. First lesson is to learn how to identify frequencies when you hear things; this can be done by buying an audio ear-training package. Second lesson is to understand that this situation was a bit different than simple feedback. The man speaking has a very sonorous, deep voice with tones that are different from other folks who often lead up front. This triggered the sound system and room acoustics, and the particular combination of all this resonated in a persistent ringing tone. The temporary solution was to attenuate around 500Hz (about 4dB or so) until the ringing stopped that morning. You don't want to leave this EQ setting in place, though, since it will make the other people who use that channel sound thin. Save and recall EQ settings for each person; store snapshots or just write it down. A more permanent solution is to ring out the system and see if 500Hz is a problem waiting to surface. If so, use your system EQ to attenuate slightly in that range.

I keep hearing feedback, but can't find it anywhere.

Many years ago I was going crazy for a few weeks tracking down a persistent, high-pitched ring. I tried everything with no luck until I happened to mention it to a friend who sat somewhere on the other side of the room. "Oh, that's Ed's hearing aid. Every so often one of us will lean over and yell at him to turn it down." When someone using a hearing aid cannot hear, they crank up the volume. At some point the device begins to actually feedback on itself, meaning the mic picking up external sounds also captures the amplified sound going into the ear canal. Amazing.

I can't get enough level on the mic.

As you turn up the mic trim you may not get enough signal on the meters, even if you go all the way up with the mic preamp. There are a couple of reasons this can happen.

- The musician is not playing or singing at full volume. Often a musician will softly talk or hum into the mic or lightly play their instrument since they're not in the full swing of the tune. This is a much lower sound level than you'll get when they actually begin playing, so try to have them play normal—ask them to run through a song if possible.

- Dynamic mics have a lower signal output than condenser mics. See the section on microphones for more on mic characteristics, but for now just remember that some mics aren't as "loud" as others. If you put an AKG 414 condenser on an acoustic guitar, then decide to replace it with a Shure SM57 dynamic, you'll have to crank up the 57 higher to get a decent level. (The reverse is also important: if you have a dynamic mic on a source, then replace it with a condenser without changing levels, you'll likely get a nasty surprise as your speaker cones shoot across the room...well, maybe something like that.)

- The mic may need to be closer to the sound source. Quiet sources such as acoustic guitars with nylon strings and soft-spoken singers don't put out a lot of sound, so you need to move the mic closer. It's better to get closer than cranking all your gain controls wide open.

- Make sure your channel faders are up around unity (0 or U). Running faders really low actually reduces signal level considerably, so keep 'em up.

It sounds muddy and boomy.

There are a few reasons why your sounds are muddy or boomy in the low end.

Reason #1:

With certain types of microphones, close placement to the sound source results in a low-frequency boost called *proximity effect*. This occurs with directional mics such as cardioid (uni-directional), hyper-cardioid, super-cardioid, and bi-directional. Omni-directional mics are not affected by this. An example would be an acoustic guitar that sounds unnaturally boomy when close-mic'd.

Solution #1:

Move the microphone a little farther away from the source. This will attenuate (reduce) the low-frequency boost.

Audio example 55: Moving the mic back to reduce boominess

Solution #2:

Turn on the low-cut filter on the microphone (if it has one) or on the console channel the mic is plugged into. This will sharply attenuate all low frequencies below a set point, usually around 75 or 80 Hertz. Digital consoles have an adjustable filter so you can set it differently for each instrument. Of course, for instruments that have low-frequencies, such as upright bass and kick drum, you will begin to lose the sound of the instrument itself, so we don't typically do this on those parts.

Audio example 56: Using a low-cut filter to reduce boominess

Reason #2:

All acoustic sound sources have a resonant frequency range. This is a region of frequencies that stands out stronger than the others, making it sound muddy. Resonant frequencies are usually in the low-mid range, around 200—400Hz. If not tamed, the combined effect from all your mics will result in an overall muddy sounding mix that lacks clarity in the bottom end. A common example is a kick drum or upright bass that has a muddy sound.

Solution:

Use a parametric EQ to find the offending frequency range.

A parametric EQ provides an extra control that sweeps around the frequency spectrum to find specific areas to boost or cut, as opposed to a graphic EQ which is set to a fixed frequency. Look at the LM (low-mid) band on the graphic here; it includes a boost control (-12/+12) and frequency select (100—2kHz).

How to find the resonance:

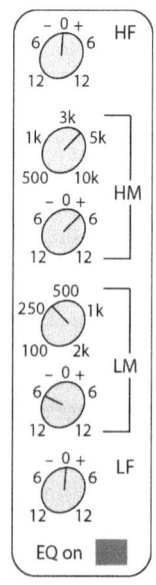

Listen to the guitar mic by soloing the channel. Turn on the EQ. Turn up the gain for the low-mid region at least 6dB or so. (dB is a unit of audio level measurement, in this case indicating the amount of level change.) Now rotate the frequency select control adjacent to it and listen for the changing sound of the frequencies as you move up and down the scale. Turn it back and forth until you can distinguish a region which stands out beyond the others and sounds quite muddy. Now reduce the boost control back to a negative number, perhaps between -3 and -6dB. Cutting too much will thin out the sound, so find a compromise.

Some parametric EQs provide one additional control that allows you to narrow the region of frequencies being cut (*bandwidth*, or Q). By narrowing the bandwidth of frequencies being affected, less of the overall sound will change, allowing you to attenuate (cut) only the offending frequencies.

This same procedure can be used to isolate other EQ problems, such as harsh hi-mid tones.

In the example below, we first boost the low-mid EQ band and sweep around looking for an especially muddy region. Then we play it with that area attenuated. Next we boost the hi-mid control and look for some nice clarity and presence, finally playing the finished EQ curve with attenuated low-mid and slightly boosted hi-mids. The settings in the graphic here are fairly close to what was done in this audio file.

> Audio example 57: EQ a guitar to clean up mud and improve clarity

Steve's vocal mic sounds fuzzy.

This is most likely distortion, not his mustache, and is caused by levels being cranked up too high. Audio systems can only handle so much signal level before they peak out and start doing nasty things. The first culprit is the mic preamp on the console. Turn this down, then nudge the channel fader up if needed to compensate. See if the meter indicators on the channel are flashing red.

What also happens is that one week you've got the soprano using mic #9, with levels set for her, and the following week the bass dude picks up the same mic. If the bass singer has a louder voice it'll come into the console much higher. The easiest solution is to assign vocal mic channels to specific singers. This allows you to fine-tune and set your signal processing, such as EQ and compression, permanently for each vocalist. You can also store individual presets for each singer on a digital console.

> Audio example 58: Distorted vs clean

Our sanctuary seems to sound like, well, a barn or warehouse.

I can relate. My church used to sound awful because there are huge walls that had nothing on them to control sound reflections. Lots of drywall, an incredibly high ceiling, and nothing to absorb or control sound except for the people sitting in the pews. A pastor once asked why congregation singing seemed so quiet. It looks like they're singing, but what happens is the sound goes up into that high ceiling never to return again, kinda like Enoch (look it up). You need something to reflect it back down in a controlled way along with reducing the reflections around the bare walls. See the room acoustics discussion later in the book to give you some background and ways to improve your given lot in life.

I read in your book earlier that spacing instruments apart helps reduce leakage into the microphones. I tried scattering them around the stage, but now the band hates my guts.

I doubt they hate your guts, but they have a valid point. Musicians must be able to see and connect with each other. Even if they're wearing headphones and can hear everything, it's tough to play together when they're not together. If you close-mic instruments and voices and use direct boxes for as much as you can, you'll be okay. Work with them to find a stage layout where they are comfortable, yet not right on top of each other.

I've got a choir mic on a stand out front and it's feeding back like crazy.

The farther out front you mic things the quicker you'll get feedback. Let's say you've got a choir or children's group standing near the front of the stage. Assuming your main speakers are overhead, chances are you're about directly underneath them at this point. Move the group back (along with the mic), or try raising the mic and pointing it downward more; the back of the mic, which provides the most rejection of incoming sound, now points directly toward the speakers. You might also move the mic closer to the group if possible.

I see the term "dB" everywhere...what is it?

Decibels are used to measure audio signals, both electronic and acoustic. A dB is not a specific quantity, but rather is used for comparison between two measured signals. It's kinda complicated, and there are different dB applications for different types of measurements: signal voltage, signal power, acoustic sound pressure, and so on. Mixing consoles use dB meters to indicate signal voltage level running through the board, whereas you can measure how loud your sanctuary mix is with an SPL meter (sound pressure level).

For the curious among you, the phone company originally used the term *Bel* to represent signal level measurements, naming it after Alexander Graham Bell. The audio industry adapted this for audio signals, but found that they needed to measure smaller increments of change. Thus the term *decibel*, which is 1/10th of a Bel. Feel better? No? Go read the section on dB later in the book.

Things to listen for with spoken word mics (Pastors, leaders)

The voice sounds distant.

The mic should be fairly close to the person. Pulpit/podium mics should be pointed toward the person, no more than a foot or two away. Make sure it's not facing sideways or backwards (you laugh...it happens). Lapel mics should be located about the size of a person's fist down from the chin. Too far down and you'll get a more distant room sound. The best solution is to use a head-worn microphone that loops over one ear and holds the mic capsule to the side of the face.

The voice sounds boomy.

Consistently boomy? Probably too close to the mic and generating proximity effect (exaggerated low frequency response). If it's a lavalier, lower it a bit. Podium mics should not be right in their face; position it a little above their mouth, or pull away several inches. Turn on the low-cut filter on the console channel. Most voices and instruments also have a muddy, resonant range in the low-mids that clouds the sound. Carefully attenuate somewhere between 200 and 400 Hz until it clears up. Too much, though, and you'll kill the bottom end of the tone.

It's difficult to understand what's being said.

As long as the volume in the room is high enough, make sure you've dealt with any low-mid muddiness as we just described. Now boost the hi-mid EQ a couple dB or so around 4kHz. This region provides presence and intelligibility. Too much, though, and it will get edgy and harsh. This also works for your singers.

Now, it could be something completely different, like when we first started attending our current church and couldn't understand a word the Pastor was saying. He looked serious and profound, so we just nodded and went with it. When I started looking into the system we discovered one of the main speaker drivers in the front cabinet was not working. At all. The others were terrible speakers that shouldn't have been installed in the first place, along with the fact that they were facing *down* instead of toward the seating area. I can't explain it. But we fixed it soon afterwards.

Every so often it goes "BOOOMMM".

This is from hard consonants such as "p" and "b". These sounds force a huge wave of air from the mouth; when it hits a microphone diaphragm it overloads it, causing a sonic explosion. Hold your hand in front of your mouth and talk. Feel those puffs? They generally travel downward, so keep your mics away from this area. Head-worn mics are designed to stay off to the side. Lavaliers should be down several inches from the mouth; these are problematic because people tend to do annoying things like, say, look around while they talk. When they look down, such as when reading something, this forces air straight into the mic. For handheld mics, don't let them stick it in their face like they're on MTV or something. Hold the mic close below the chin, underneath where the wavefront travels.

It sounds hollow, weird, from outer space or something.

This is a phasing issue where the voice sound goes directly into the microphone, but is also being reflected from a nearby surface (pulpit, music stand, Bible held in their hand). Listen when the person moves forward and back from the pulpit and you'll hear the weirdness change. This is from different frequencies being affected due to the changing distance (different frequencies have different wavelengths). Try lining the pulpit surface with something soft, like a mat or cloth. If possible, change the angle of nearby surfaces so they don't focus sound directly back into the mic. Show leaders how to read from a Bible or paper by holding it at a slightly different angle away from the mic.

But it still sounds hollow!

Make sure you don't have two microphones, such as the podium and lapel, turned on at the same time. It doesn't make it louder—it destroys the frequency response and therefore the tone. Only one at a time.

What's that static, clicking sound on the pastor's mic?

Assuming you're using a wireless mic, you've either got reception issues or a weak battery. Wireless systems include the microphone, transmitter (the little box that clips on their belt), and a receiver (this is what you connect into the console). If the receiver doesn't get clear communication from the transmitter, it might cause a drop-out in the audio. The culprit might be some steel ductwork or other metal structural components in the room, or it might be that the two units cannot "see" each other. Try adjusting the antennas on the receiver, find a clear line-of-sight path up to the stage, and move the units higher. Receivers usually have some sort of meter or indicator to show signal strength or loss, so this can help you out. Check and replace your batteries regularly; if the battery is getting low, you'll get these reception issues for sure. Buy good batteries, not the store brands at the check out counter.

Every time someone using a wireless mic stands in a certain spot, it clicks or drops out.

Wireless is not perfect by any means. You've discovered a dead spot where the receiver cannot get a clear signal from the transmitter. This is usually easy to fix by moving the antennas, raising the receiver to get a better line of sight to the stage, or moving the receiver to a different location. Train your people so that if they hear their mic stop working, just move over a step or two.

What on earth is that?

And then there are times when you don't even know what the question is. One Sunday our pastor began talking and it sounded really, really low and thin. The guys were scrambling to see what it might be, and when I walked over to take a look I tried bypassing the external EQ we had at the time on the voice mics. Sure enough, the pastor sprang to life and all was well...except I have no idea why this suddenly caused such a problem. It's possible someone changed something in the signal chain, which makes it difficult to diagnose on-the-spot. Here is the setup we used to run for voice mics when we had an analog console:

All voice mic channels (pastors, leaders) were routed to a common sub-group, not the main mix. We inserted a compressor and external EQ on this group bus. Here's the signal flow: The sub-group insert send goes to the external EQ, then from the EQ to the compressor, and from the compressor back to the bus insert return. The group is then routed to the main mix bus. That day the compressor looked like it was working normally. Who knows, but this is why knowing your equipment setup and signal flow is crucial. You *have* to know how things are wired and routed if you want any chance of solving problems.

Worship team issues

I can't hear myself in the monitors.

A broken record that will never be resolved until Jesus returns. For now, a solution depends on the system you're running. If you have the luxury of a personal monitor system, each individual can customize their own monitor mix, so this person can simply crank themselves up until they are deliriously happy (or go deaf). These usually feed headphones or earphones, so there is no acoustic sound at all. If you have floor monitors (called *wedges* because of their wedge-shape), this is more problematic because everyone, including the congregation, can hear them. If they are sharing a monitor with, say, the entire vocal team, then that's a real challenge. The vocal group has to compromise on a mix where they each can hear what they need. It's great fun for me to watch a singer casually reach over to turn their channel up, then a bit later another singer reaches for their knob, and later the first person tries a bit more, and so on.

I can't hear the piano.

Singers need two things to sing by: timing and intonation (pitch). Drop the kazoo and add more piano in their monitor mix. See what else you can reduce that might be cluttering the monitors. You also need to go up on stage and listen yourself to really understand what they're dealing with. You might have that piano aux send cranked to the max for the monitors, but standing beside the singers you hear nothing but drums...because the set is located right beside them. Don't assume anything—go check it out yourself.

I can't hear myself out in the sanctuary (says the bass player).

Your musicians should not be relying on the house mix to hear what they're playing. But, even though they are using wedges or headphones, there's something reassuring about hearing themselves in the room. Sometimes they are correct—their channel is turned off or down too low. Other times the mix is fine. Help them to understand that with all they get in their monitors, it might be tough to hear much of the sanctuary mix. Trust goes a long ways, so the more you show your team members that you are doing your best, the less they will question and wonder.

Congregational complaints

It's too loud!

Maybe it really is. Most church services are not rock concerts, so you have to think differently. Sound systems exist to support what's there, not take over everything.

Maybe this person is sitting close to a loudspeaker, musician monitor, or the brass section. Suggest a better location for them to sit, and consider removing seating if possible in areas prone to this problem.

I can't hear _____

Again, it might depend on where they are sitting. You might have a dead spot in the sanctuary which requires a system adjustment or modification. Maybe something's wrong, like a mic connection, or more serious such as a main speaker that's not working. It might also just be turned down too much overall. It's difficult to find the right volume balance to make everybody happy, so seek input from several people who sit in different sections of the sanctuary and see if you can get around the room during a service to hear for yourself.

If they can't hear a particular part, maybe the fader is too low on that channel or the source is actually not playing or singing loudly enough. Our vocal mix varies all the time, even from song to song, if a singer or two doesn't know a particular song very well. We sometimes allow newbies to play along with the team (though only during rehearsals), but we don't run their signal into the house mix. And there are people who are related or in love with one of your team members and only want to hear them above all else. I'll let you figure how to deal with this one.

It hurts my ears.

This could be from high volume, but it could also be an EQ problem, meaning that certain frequencies in the sound might be too harsh and overbearing. I've heard this many times where the high or high-mid frequencies are turned up too much. The idea is to help the vocals "cut through", but when the ears begin bleeding I think we need to dull the edge a bit. Typically frequencies in the 1-2kHz and 5-7kHz range can be uncomfortable when turned up too much.

Drums are straight from Satan's toybox.

Okay, you got us. This is not an audio problem, so redirect this comment to your pastor for a scriptural rebuttal. Just so you know...there is no scripture that prohibits any particular instrument in worship, and I do believe there are several references to "gongs and cymbals". However, it might be kinda interesting to know what's in Satan's toybox, though I'd rather not go visit.

Enough already…just help me fix my guitar sound!

Ok, ok…we've covered a lot of ground. Here's a quick summary to get you going this week. Keep in mind every situation will be different: instruments, players, gear, style of music, your PA system, etc. You can start with this, but you probably need to tweak it a bit to match what you're getting into the console. By the way, use headphones for fine-tune listening; you're probably not going to hear some of this over your main PA. Just make sure you do your final shaping through the system so you get it right in the room.

Bass guitar

Plug it into a quality, active direct box. If you get a hum, flip the ground lift switch. With a digital console, turn on the channel compressor: 3:1 ratio, attack set to medium-fast, medium-long release, and set the threshold so it's flashing 3-4dB reduction—not continuous. If it sounds a bit muddy or boomy, fire up the EQ and attenuate low-mids several dB (somewhere between 200 and 400) with a fairly narrow bandwidth. Add some presence and articulation with a wide-band 4-6dB boost @2–3k.

Here's a bass track that starts raw, then adds an EQ with a narrow-band 7dB dip around 213Hz, a wide-band 6dB boost @2k (for presence and definition), then a couple dB boost at 61Hz for a bit more bottom. The bass sound itself was pretty round and dark; it sounds good, but there's only so much presence and bite you're going to get. We then add a compressor with medium-fast attack, medium-long release, and 3:1 ratio. Set the threshold to get it flashing a bit. The goal is to even out the performance and tighten the sound so it's not jumping all over the place.

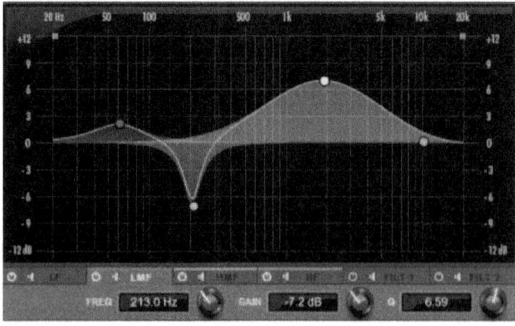

Audio example 59: Bass guitar (raw + EQ + Comp)

Acoustic guitar

Use the pickup unless they scream at you and plug it into a quality, active direct box. Flip the ground lift switch to whatever makes the hum lower. Set your compressor at 3:1, not-quite-fast attack, medium release, fairly high threshold so it's flashing 2-3dB reduction. You don't want a lot on here unless it's an aggressive background rhythm instrument (if so, bring threshold down a bit and maybe bump the ratio up to 4:1). Drop some low-mids with a narrow-band attenuation and boost wide 3-6dB around 3-5k for presence and bite. Add a low-cut filter around 75Hz to get rid of extraneous low frequencies that will muddy the mix. If you're running reverb, maybe a small amount will give it some space in the mix.

This example has a narrow 4dB dip at 380Hz and a simple high frequency shelving boost that starts low around 1k. This opens up the entire upper-mids and highs for more presence and air. Notice the low-cut filter at 50Hz. The compressor is set for a gentle action at 3:1 ratio, medium attack and release, and just enough threshold so it's flashing 3-4dB—just enough to keep it in its place in the mix.

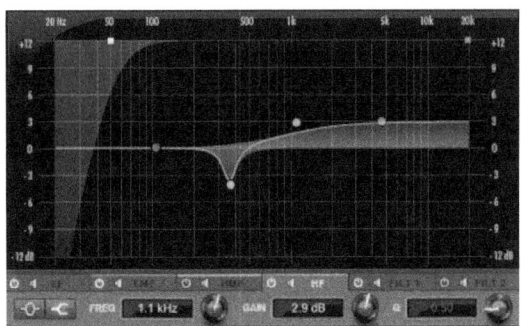

Audio example 60: Acoustic guitar (raw + EQ + Comp)

Electric guitar

Take the signal straight from their guitar, pedal effects board, or amp simulator into an active direct box. If they're running an amp, point a Shure SM57 at the grill about an inch away, mid-way out from the center of the speaker cone. Set the compressor a bit more aggressive than for an acoustic: 4:1 ratio, lower threshold until it flashes 4-6dB reduction, fast attack and medium release. Maybe drop some low-mids a bit to clean it up, and if it's not too harsh already add a couple dB around 3-4k for presence and bite. A low-cut filter around 75Hz will eliminate stage rumble that muddies your mix.

Here we've got an SM57 close on an amp. EQ has a fairly narrow dip around 480 to get rid of a honky, nasal sound, a slight bump in the bottom, and a broad boost centered at 2.3k. A fairly aggressive compressor running at 4:1 with a decent amount of gain reduction helps tighten the guitar to fit in your mix.

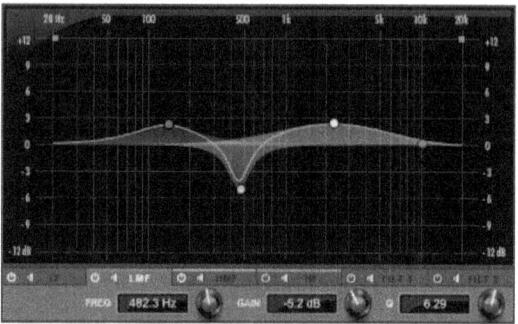

Audio example 61: Electric guitar (raw + EQ + Comp)

Here's what the stuff we pulled out of the low-mids sounds like on its own.

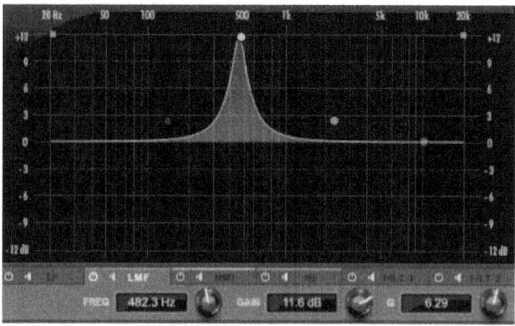

Audio example 62: Electric guitar 480Hz yuck

Piano

For grand pianos, tape a PZM under the lid facing toward the higher-frequency strings (the shorter ones); put the lid on short-stick or fully closed. You'll certainly need some EQ, so look for low-mid attenuation to reduce muddiness and add some presence and brightness with a 3-4dB boost at 4k and a small shelving boost around 8-10k. Even on a piano, turn on the low-cut filter around 50-75Hz to reduce rumble. The exact amount depends on the piano itself and the overall sound system. You can run a compressor, but at a low setting: 3:1, not-quite-fast attack and medium release, fairly high threshold so it only flashes a couple dB.

Our example here has a narrow 5dB dip at 427Hz to reduce a nasal, cloudy sound, a slight bump at 93 for bottom, and a sweeping high frequency shelving boost starting low at 1.4k. This captures nice presence in the mids as well as the sparkle on top.

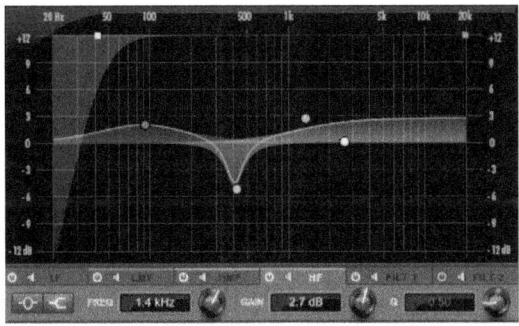

Audio example 63: Piano (Raw + EQ + Comp)

Drums

Put a large-diaphragm dynamic mic inside the kick drum about halfway, pointed toward the beater. If there's a front head with a hole, stick it in the hole as far as you can. If there's no hole, tell the drummer to go home. Put a pillow or blanket inside the drum to dampen the vibrations. For snare, put a small-diaphragm dynamic inside the edge of the rim about two inches above the head. Same for mounted toms. Floor tom is best with a large-diaphragm dynamic or condenser a few inches over the head near the edge. For cymbals, place two condensers either spaced apart over the kit (left and right, pointing straight down) or in XY position about 3 feet over the kit. Each drum will need some narrow-band low-mid attenuation (less so for toms) and presence boost around 2-3k. Kick has a nice meaty sound around 80Hz that you can boost a bit. Use a low-cut filter on snare, mounted toms, and overheads set at 80-100Hz. I like to compress the kick a bit, more so with the snare, maybe a tad on the overheads (cymbals). 3:1 ratio, 2-3dB reduction, not-quite-fast attack and medium release. You want to tighten it up and add punch, but not squash it too much, so move the threshold up if it's hitting too hard. Here are kick, snare, and mounted tom examples.

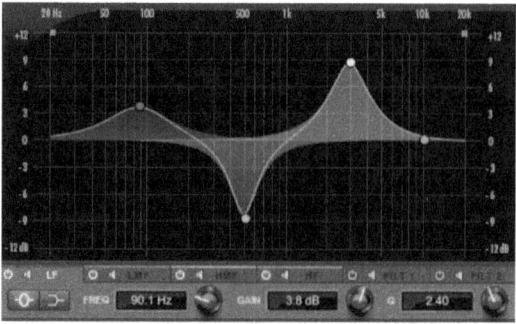

Audio example 64: Kick (Raw + EQ + Comp)

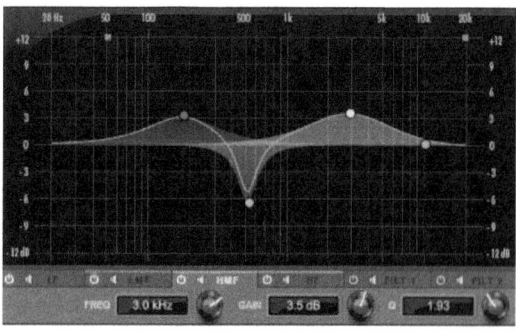

Audio example 65: Snare (Raw + EQ + Comp)

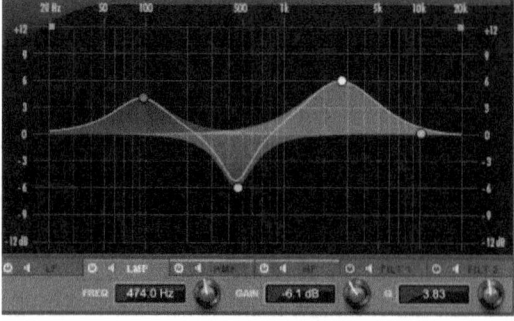

Audio example 66: Tom (Raw + EQ + Comp)

Vocals

Use dynamic mics and have them hold it close just under their mouth—not in front. Try to keep their monitor levels as low as possible to reduce feedback as well as stage sound that leaks out into the sanctuary. A small amount of narrow-band low-mid attenuation will clean up any cloudiness, and a wide-band 3-4dB boost around 4k will add presence in the mix. Exact frequencies and amount of boost/cut will vary between singers, so you'll have to store settings for each person. Use a 100Hz low-cut filter on all your vocals. A bit of compression can control fluctuating dynamics, so start with 3:1, fairly fast attack and medium release, and set threshold for 2-3dB reduction. Again, you'll have to adjust this for individual singers as some have really strong voices, others more timid. If you dial it in right, you'll have a tighter ensemble sound. A touch of reverb with a short reverb time (1 or 1.2 seconds) will add a sense of space while helping mesh the vocals together. You can even experiment with a low-level delay in the background. Just remember to turn these off when they're talking between songs.

Audio example 67: Vocal (Raw + EQ + Comp)

Pastor

For talkie mics, you're looking for a fairly consistent volume level that's clear to understand. Start with a quality mic as this makes all the difference in the world, preferably a head-worn model, which is far superior to lavaliers. A touch of high-mid boost can enhance clarity, and you might need a little low-mid attenuation to reduce cloudiness. Turn on the low-cut filter at 100Hz. Set your compressor for 2-4dB reduction, fairly high threshold. The idea is to control the really loud moments, but not be acting continuously. If you're feeding a recording system, put a different compressor on that bus with a much more aggressive setting: at least 6dB continuous reduction (higher ratio, lower threshold). Different individuals will need their own EQ settings (and possibly compressor settings as well), so store these and load them into the channels as needed.

Overall notes

Remember, compressor settings are only as good as the incoming level, meaning if you turn the mic preamp up or down, it'll hit the compressor harder or lighter. This has a big impact on your sound, so just be aware of signal flow and how everything effects each other.

EQ settings depend on your overall system and sanctuary sound. A piano in one room might need 6dB boost at 4k to get enough bite and presence, whereas this might come across as too harsh in a different church. If you re-tune your sound system, then all of your channel settings will need to be checked again.

Try to help your musicians fix the sound at their instrument. Drum heads and tuning, old guitar strings, and cheap guitars have everything to do with the quality sound you're able to get in the mix.

Rotating team members and pastors will require their own EQ and compressor settings. Store these and remember to recall each week.

A great many of you are running analog consoles, which means you don't have compressors on each channel and you can't store EQ settings. Don't despair—you can still get a great sound. The major factors are the musicians, their instruments, your miking technique, the sound system, and the room acoustics. The only really crucial need for a compressor is to record your sermons, and you can buy an outboard compressor just for this. To remember EQ settings, just do what we did in the studio decades ago before automation—we wrote it down and took pictures.

Next steps

My church system has a rig that allows me to run software plug-in processors on my console channels. If you've done any recording with Pro Tools or something similar, you're familiar with having different types of EQs, compressors, and so on. This is great because now you not only have a compressor for each channel, you've got different flavors of compressors to choose from. And yes, it makes a difference. Take some time and try different models on your acoustic guitar and pick the one that makes it shine. I use a chain of processors on various channels that goes something like this: surgical EQ (for attenuating muddiness, low-cut filter), compressor, shaping EQ (add presence, sparkle, maybe low end), and perhaps a final leveler/limiter that helps fit the instrument into the mix. I absolutely love multi-band compressors that can be set to compress at specific frequency ranges rather than the entire sound. For example, I have a couple vocalists who sing really well, but can get a bit edgy in the upper-mids. Instead of changing the entire sound with an EQ, the multiband will only kick in when things get too much in that range while leaving the other regions alone. Very smooth.

I should also point out that at some point you might just be chasing a fantasy. Wonder why your mixes still don't sound so great even though you've memorized and practiced everything in this book? It's probably your musicians. A majority of churches don't have the luxury of professional or even *good* musicians. If your players and singers can't play and sing, and if they have poor instruments and gear, then you can't really fix that. Now I wouldn't go around telling everybody, but just realize you can only go so far. Do what you can to improve yourself: learn more, practice, and find a mentor to guide you.

These ideas will get you pretty close. At that point it's up to your ears to know how to balance everything and tweak these settings. Experienced live engineers will go way beyond all this and do all kinds of layering, effects, and so forth to really fine-tune their mixes, but don't worry about that. If you can take what we've described up to now and get a handle on it, you've improved your skills and your congregation's experience a huge amount.

LEVEL 4

Digging deeper

This section provides more detail and background explanation for some of the equipment and procedures we've been working on throughout the book. You don't need all of this stuff to run sound on Sunday mornings, but it can make a big difference in helping you do a better job. There are other books out there that go even deeper, so instead of me trying to replicate all of that (and make the book agonizingly longer), I would encourage you to add them to your library as you find time.

Mixing consoles

When prospective students and their families visit the college where I work, we show them our studio facilities. One look at the large recording consoles prompts them to ask "how do you learn what all those buttons do?" Mixers can be intimidating, but they can be broken down into sections and functions. If you've got some experience, you're most likely  beyond that, and so instead of seeing a mass of buttons and controls, you're beginning to see *patterns*. This means that you see *operations*, not controls; to run a mic signal to the mix your mind automatically filters out all the other components except for what's needed to route the signal. Remember this graphic from the very beginning of the book?

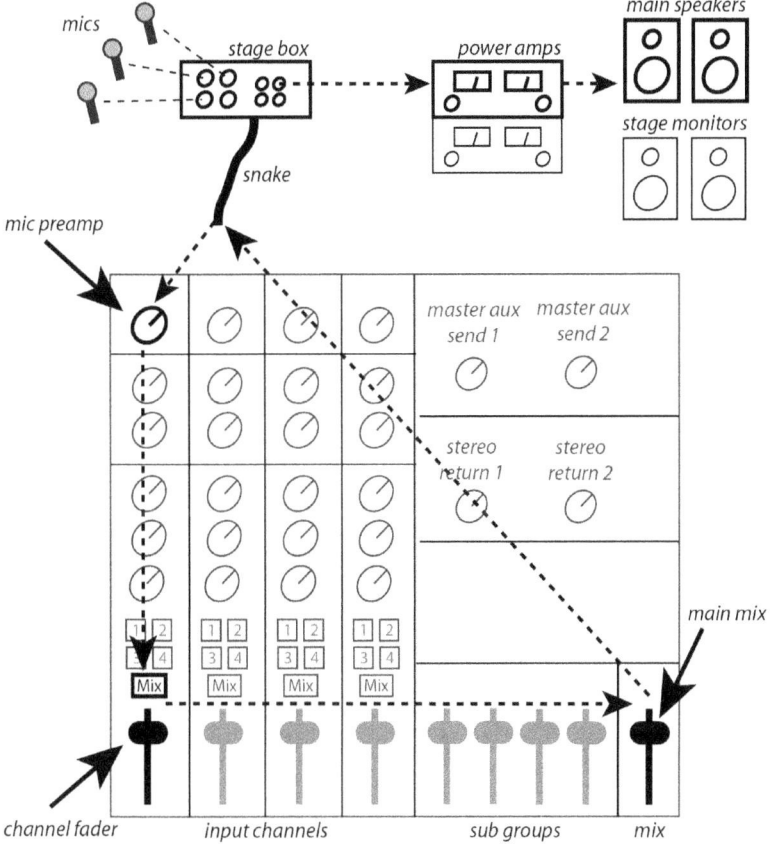

mics / stage box / power amps / main speakers / stage monitors / snake / mic preamp / master aux send 1 / master aux send 2 / stereo return 1 / stereo return 2 / main mix / Mix / channel fader / input channels / sub groups / mix

This graphic highlights only those controls needed for a particular operation, in this case routing a mic signal to the main mix. All other items are dimmed, exactly what happens in your mind as you work at the console. Are you really good at chess? I'm not, so I look at the game board, see two rows of various medieval miniatures, and try to remember the rules for how each type can move. An expert chess player takes a quick glance and sees patterns of strategies, including not just the next move, but several moves down the road. Experts think in terms of operations and patterns, exactly what you will develop over time at the console...but only if you understand how these operations work and practice at it. The value is that it helps you operate more efficiently, be able to solve problems as they arise, and know how to figure out new situations or needs. Consider a scenario where you can't get a mic working. Instead of looking at each row of knobs and buttons, you quickly scan the controls directly related to routing a mic signal to the mix bus. You can instantly tell if you missed a switch somewhere—very quick and efficient.

By the way, although digital consoles have become standard these days, they follow the same operational principles as analog consoles. The trick is that analog functions, such as channel controls and signal routing, are mostly "hidden" in menus and touch screens. It's actually much easier for people to learn how to use analog consoles, then transition to digital. I'll lay things out analog here; once you know what you need to do, you can find the appropriate controls on whatever console you are using. Now, let's go through most of the controls and functions found on live sound mixers.

Consoles can be divided into three main components: input, output, and monitoring. Mixers take incoming signals, such as microphones, computer audio, and iPods, and send them places like the main speakers, monitors on stage, effects units, recording devices, hearing assistance systems, website streams, and so on. While this is happening, you can tap in to listen to any of this by selecting a single channel, output, or the entire mix.

Mixers come in different sizes based on the number of input *channels* and output *buses*. If you have a 24-channel mixer, you can connect up to 24 different microphones, each running into its own channel. What's a channel? These are the vertical columns of controls that are repeated over and over across the console. Each channel has controls that adjust the signal's level (*mic preamp, fader*), change its tone (*equalizer*), *pan* it left and right if you're running a stereo mix, send (*bus*) it to stage monitors and recording systems, and route it over to your main mix fader (*mix bus*) which feeds your main speaker system. If your mixer has 4 buses, that means it has four subgroups where you can combine several channel signals, then as a group they go on to the main stereo mix. I'll talk more about that later, but the idea is that these can be used to create a sub-mix of vocals so you can control all your singers using one fader, rather than trying to grab them all at once with several fingers. Small and mid-size consoles usually have 4 or 8 subgroups.

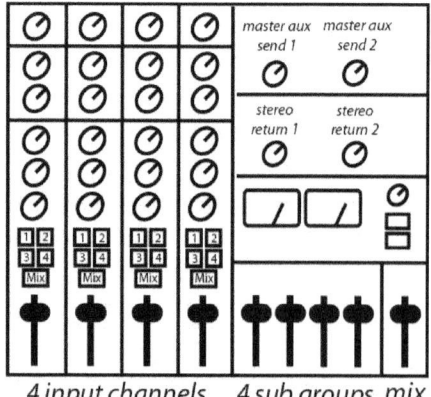

4 input channels   4 sub groups mix

Input channels

Input level

At the top of the channel you'll find the first important control to understand—the *microphone preamplifier*. Microphones have a very low signal level output, so the console needs to crank this up so it can work with it. The mic pre is a special amplifier that provides a lot of boost to the incoming signal before passing it along to the rest of the channel path. It influences the quality of your sound far more than anything else in the console, so this is a key feature that distinguishes cheap and expensive mixers.

In this same section you'll find the phantom power button (+48V) which is needed for condenser microphones and active direct boxes. You might have an attenuation pad; this reduces incoming level so it doesn't overload the channel. You'll rarely use this, but drum mics are a typical example. Finally, the "ø" symbol will flip polarity on an incoming signal, such as when you've got a mic cable whose leads are reversed (pins 2 and 3) or if you're miking the top and bottom of a snare drum (flip the polarity on the bottom mic).

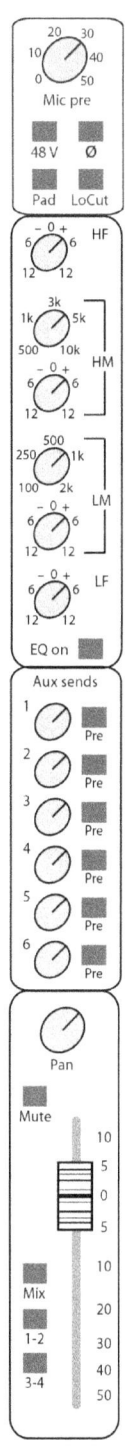

Equalization

EQ is used to change the tone of a sound. We use it to clean up a muddy sound, brighten a dull sound, or just make it sound better. You'll see labels such as *low, high,* and *mid.* EQs offer different *bands,* or ranges of frequencies, that can be controlled individually. A simple example of this is the bass and treble control on your stereo system—two bands. You'll also see a *low-cut* or *high-pass* switch, sometimes indicated by a symbol that looks like a division sign. This control eliminates very low frequency sounds, such as rumble, air conditioners, leakage from the kick drum, and nearby trucks zooming down the highway. See the EQ discussion back in *Level 3* for details on how these work and how to use them.

Auxiliary sends

All mixers have a number of rotary controls (*pots*) that simply take a copy of the signal running through the channel and send it somewhere. For a live sound situation, aux sends are primarily used to send mic signals to the stage monitor system so the performers can hear everything going on, but they are also used to send copies of these signals to *effects devices* to add reverb to your mix. Aux sends have nothing to do with getting mic signals to your main mix, so you can ignore them while you get sound up and running. The *pre* button tells the aux send to get its signal *before* the channel fader, so when you move this fader up and down it won't change the level going to the aux send. This allows you to set up a monitor mix just the way they like it (yeah, good luck with that), and move your faders as much as you want.

Fader and pan

The fader changes the level of the signal in that channel. This is the main control you'll use to balance everything while you're mixing. See the area not quite at the top where it's labeled either "U", "0", or something similar? That's the ideal point, so start there and move up or down as your mix requires. If your system is set up for stereo, then you can use the pan pot to move that vocal or instrument left to right. Go ahead, try it...it's really simple.

By the way, "U" stands for *unity,* which is the point where an amplifier (yes, faders are actually small amps) does not increase or decrease the signal—it merely passes it along without affecting it. When you push a fader higher than unity it amplifies the signal, whereas pulling it down below unity decreases signal level. We like to minimize the amount of amplification that occurs on a signal, so that's why we start at unity and try not to stray too far.

Mute, solo, assignment switches

You can turn each channel on or off using the *mute* switch (or *on* switch on some mixers). The *solo* switch singles that channel out in the monitor section so you can listen to just that mic to see who's coughing up Krispy Kremes. See the numbered switches either near the fader or at the top of your mixer? These are assignment switches, and they route the incoming mic signal either to a subgroup or to the main mix. If you don't select this, your microphone signal grounds to a halt and goes nowhere.

Master section of the console

Now, in the center or right side of your console you'll find controls for overall functions. This includes the main mix fader, any subgroup faders, signal meters, monitor source select and volume (so you can listen to any part of the console over headphones), master auxiliary send outputs, and the auxiliary or stereo returns.

Stereo main mix bus

This is where everything goes that you want heard in the main sanctuary speakers. All the individual mic signals coming into different channels are combined (mixed) here, then routed to your amplifiers and loudspeaker system. Set the fader at the "U" or "0" point and leave it there. Do all of your level/volume adjustments with the individual channel faders.

Output buses (sub-groups)

Sounds technical, but a bus is simply a wire that collects audio signals and takes them somewhere. Think of the downtown bus where people get on at different stops, then everybody jumps off at the post office. The assignment switches on the input channels put the mic signals onto a specific bus. So, if you push the *Main Mix* button on a channel, that signal goes to the main mix fader, which then feeds the main speakers in the sanctuary. Now, if the mixer has subgroups, you can assign multiple mic signals to one of these where

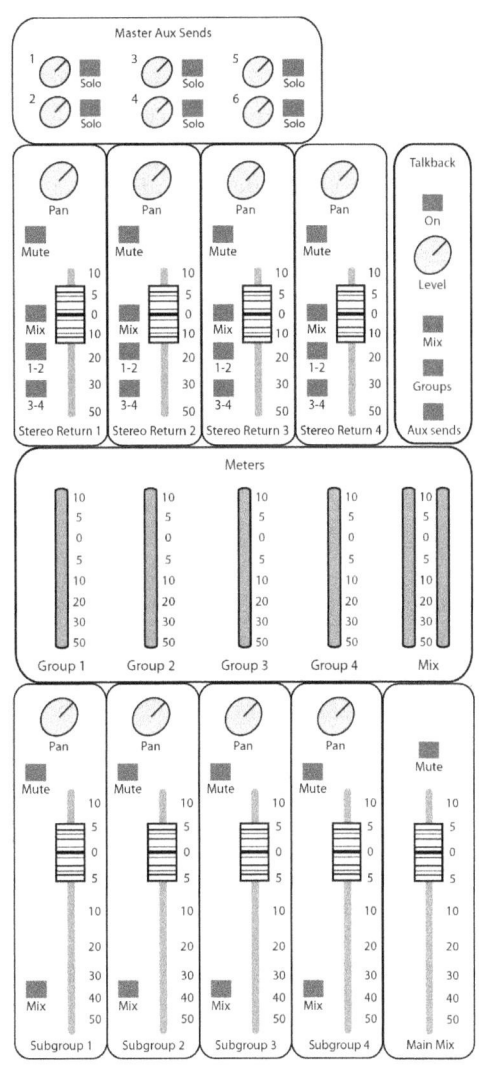

they are combined together. You can use that output bus subgroup fader to control their level, and they can then be routed to the main mix bus or even a recording system.

Auxiliary sends and returns

Master aux send outputs correspond to those numbered aux sends on every channel. You notice that every channel has an aux 1 control, right? All of these aux 1 sends feed the master aux 1 output in this master section. Whatever you turn up on the individual channels gets combined at this master output, which can then be routed to a stage monitor, reverb device, or a recording system.

Aux returns are not connected to aux sends; instead they are merely extra inputs to the console where you can plug stuff in and have it feed the main mix bus (and therefore be heard in your overall mix). So what? Once you send a group of mic signals to an effects device and select a reverb sound, you need to get that reverb sound from the device back into your mix. Take the output of the effects device and connect it to your console's aux returns (or stereo returns—same thing). Turn up that aux return level and it will feed reverb to the mix bus and make your vocalists so happy they'll weep.

You can also connect playback devices such as iPods and CD players into auxiliary or stereo returns. There's nothing magic about aux returns—all they do is provide an input to the console that feeds the main mix.

Signal meters

Signal meters are there to make sure you're not overloading the system or running too low. Try to have your signal levels through the channels and main mix bus where the meters are averaging "0dB". It's okay if they run into the red a bit, but don't let them slam the top continuously.

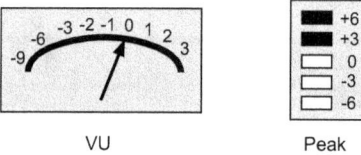

VU   Peak

Each channel will also have a green or yellow LED that lights up when signal is present (a good thing), and also a red LED that lights up when the signal is too high and overloading the channel (a bad thing). If you see the red one flashing, turn down the mic preamp level at the top of the channel.

Digital consoles

At this point you may be wondering "This stuff is great, but we have this cool-looking digital console that doesn't look much like your pictures."

Don't worry, you haven't wasted your time. Digital consoles do the same thing as analog boards. Basic concepts of channels, bussing, aux sends, inserts, and so on carry over; it just looks different. The trick is to make the transition and find what you need on a digital mixer.

Advantages of digital consoles include processing for every channel and bus (compression, EQ, gating), built-in effects such as reverb, storage of scenes and presets for easy recall, and the ability to lock out folks who shouldn't be messing with your setup parameters. You save lots of money by not buying compressors, EQs, and effects units to fill up a rack next to the console. They connect directly to your computer, making it easy to record, run the board from the computer, or even incorporate iPads and other mobile devices to control the console from anywhere in the room.

The potential downsides? There's more to break on a digital console, and much of this means the entire board goes belly up. They can also be fairly complicated to learn and understand. Most current audio people grew up with analog gear, so they have a head start of what to look for on digital boards. Learning audio for the first time is difficult with a digital system, which is why I presented everything in this book as analog. Anyhow, let's take a quick look at how to compare analog and digital consoles from an operational standpoint.

Here's an analog console surface. You can physically see every channel, EQ, fader, bus assign, and aux control.

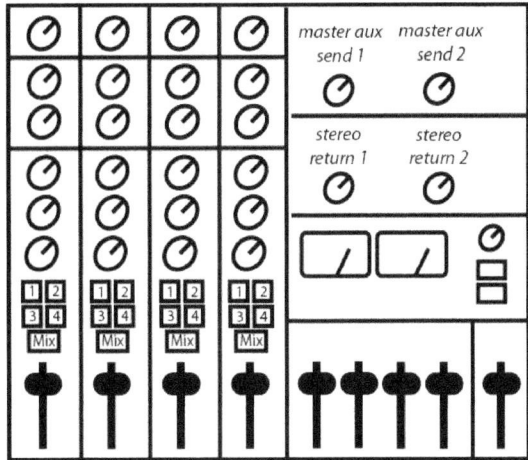

And here's what it might look like on a digital mixer. You still see channel faders, a mic trim, and maybe a couple other controls, but many functions are now hidden under menus or controlled in a master section. EQ, aux sends, busing, and other processing are available on each channel, but you first select the channel number and then adjust the parameters in a central window. Want to EQ the next vocal? Select that channel, then use the same controls for EQ.

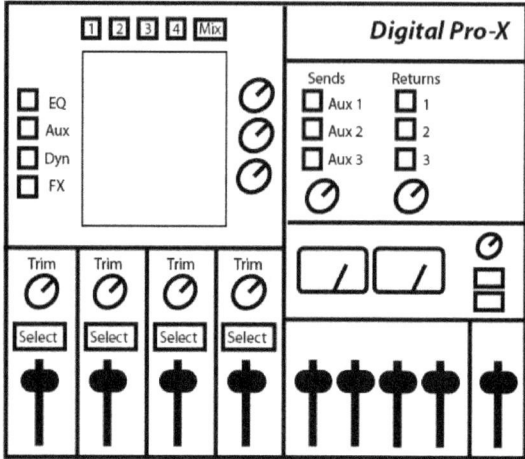

The trick to all this is understanding the operations you want to perform and what's involved. No matter what kind of board you have, you always need to bring a mic signal into a channel and route it to the mix bus. There will always be situations where you want to bus vocals or drums to a sub-group. Stage monitor mixes must be set up using pre-fader aux sends. All you have to do is find out how to get it done on your particular console.

Cables

There are two ways to get an audio signal from one place to another—broadcast it wirelessly or run it through a cable. Cables are by far more reliable and trouble-free than wireless systems, so use a cable whenever you can. There are lots of different cables in the world, but you only need a few in order to operate your church system.

Microphone cables

Microphone cables use XLR connectors, male on one end, female on the other. These feature three pins that carry signal and a ground. If you look really closely, you'll notice that the pins are numbered 1, 2, and 3. If you purchase all of your cables, don't worry with this, but if you make your own you need to make sure that you always connect pin 1 on one end to pin 1 at the other end, and so on. Take care of your cables.
They're expensive, and when people step on them, twist them, or use them to lasso toddlers the tiny wires inside the rubber jackets begin to break and eventually fail. To plug in your mic, insert the connector until you hear it click into place. Now, to disconnect it make sure you push the little release button found either on the cable connector or on the panel.

Guitar cables

Electronic instruments use what are generically called "guitar cables". They have 1/4" connectors on each end, with one signal wire inside surrounded by a braided shield for ground. Use them for guitars, bass, keyboards, amp simulators, etc. For keyboards, plug it into the main or mono output (not the headphone jack); guitars only have one jack, so you can't go wrong here. Plug these into a direct box, which is then connected to your mic stage panel with a standard microphone cable.

Just so you know, guitar cables sometimes act like an antenna for outside interference (RF). I once saw a guitar player plug into his amp and lo and behold, a local radio station started broadcasting right there in the studio! They are generally okay as long as the cable is well-made and not too long. For most studio cabling, however, this is not acceptable; I'll explain that next.

Free tip: if you get a hum after plugging in a guitar or keyboard, flip the ground lift switch on the direct box. It should either go away completely or at least get quieter.

Balanced 1/4" cables

Guitar cables are fine for guitars and stuff, but they don't cut it for system connections. The shield wrapped around the signal wire doesn't always prevent outside interference from getting in. For professional audio we use *balanced* cabling. Instead of the single wire used for  guitar cables, balanced cabling has two signal wires (+ -) and a shield. This is why you see three pins on XLR microphone cables. The trick is to run the two signal wires out of polarity with each other, meaning the + and - signals are reversed. The equipment specifically looks for this inverse relationship and says "pass, friend". Outside interference trying to jump onto the highway lands on both signal wires identically without this inverse configuration. When this intruder arrives at the input of an audio device, the equipment notices that it's not out-of-polarity and barks "halt or I'll shoot".

What this means to you, however, is that you must use balanced cables when you're connecting various devices such as reverb units and amplifiers to the console. This includes XLR and certain 1/4" cables that have an extra ring around the jack. An unbalanced 1/4" cable has one ring, a balanced 1/4" has two.

Insert cables

These special cables are used for connecting external compressors and noise gates into a single channel or bus. They feature a single, stereo/balanced 1/4" plug on one end that splits to two 1/4" cables at the other end. The stereo jack has nothing to do with stereo sound; it merely allows the manufacturer to combine the signal send from the console with a separate return signal from 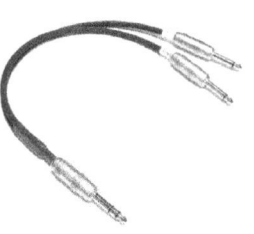 the processor. Why use a split, rather than two separate jacks? Money and space—it's cheaper to install a single jack on each channel, and it saves space on the console. Plug the single end into the insert jack, and the other two ends into the input and output of the external device. Note that these connectors will be labeled as *send* and *return*, so it matters which one goes where.

iPod and computer audio

The tiny wire used to connect a music player or computer to your headphones is an 1/8" cable (go ahead...measure it yourself). All personal audio devices use this type of cable, but it's not all that common on professional audio gear such as your console. We often need an adaptor to bridge the 1/8" to 1/4" or perhaps to a trusty old RCA, depending on what inputs your console has. Note that the 1/8" cable is most always stereo (you'll see a second black ring around the plug), so either you use an adaptor that splits this into two 1/4" connectors, or into a single 1/4", thereby providing only one channel of audio.

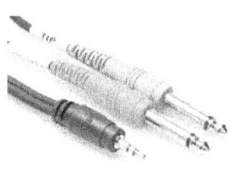

A great solution is to purchase an audio/video direct box designed to accept an 1/8" input and convert this to an XLR output (among other things). I use one at church near the piano which allows me to plug in my iPhone and play examples during rehearsals. The output feeds a regular XLR mic cable that's connected into our stage snake and then into our stage monitor system. The sound guys are deliriously happy with this arrangement since I don't have to yell "go to the second verse" or "play the ending one more time".

CD players

For years, devices like CD players, cassette decks, VCRs, and DVD players used RCA cables. Also known as *phono* cables, these were the most common type found in home entertainment systems. If you want to connect a CD player to the console you might need an RCA to 1/4" adaptor, like I described above for

iPods. If preserving a stereo signal is important (such as for splitting vocal demo tracks on song track CDs) then split it to two 1/4" connectors and plug into two channels on the console (or into a stereo return).

Digital audio cables

One of the most common cables used to transmit a digital audio signal is the optical Toslink. You'll find these on home A/V gear and most professional audio devices. A single cable carries a stereo or surround-encoded signal, and they provide a much higher quality sound than analog cables. Coax digital cables (which look like RCAs) do the same thing, but they're not as common as opticals. And then there is HDMI, which is

the best of all of these. HDMI carries high definition digital audio and video and has become the standard for home theater (and common in professional applications as well).

Speaker cables

Speaker cables connect the outputs of amplifiers to your speakers. They're thicker than other audio cables because the wires inside must be heavier to handle the higher current required for driving speakers. If you know anything about the electrical wiring in your house, wire that is too thin can overheat and cause a fire if the current load is high. Likewise, audio cables are categorized according to gauge, which refers to how thick it is and what kind of signal load it can handle. Strangely, higher gauge numbers go with smaller cable; regular mic cable is typically 22ga, while speaker cable could be 12 or 14ga.

You'll find a variety of connector types such as 1/4" (like a guitar cable), Neutrik's SpeakOn, or banana plugs (honest—see for yourself). Unlike all the other balanced audio cables that feature two signal wires and a ground shield, speaker lines use two wires only, labeled + and - (or red/black). It doesn't matter whether you use red or black for +, just be consistent. If red is connected to + on one end, make sure you use red for + at the other end of the cable.

Neutrik SpeakOn   *Banana plug*

Multichannel stage snakes

For decades the only way to get multiple microphone channels from the stage to your console was through a heavy snake. Take several mic cables, bundle them together with a thick rubber jacket, put individual connectors at one end and a mic connector panel box on the other, and you've got a standard-issue analog snake. Many snakes also provide a few channels of line-returns, meaning they have 1/4" connectors and are intended to send signals back to the stage. These are used for sending your main mix, aux monitor mixes, and other signals to your amps, monitor system, etc.

The downside to analog snakes? They're heavy and bulky. As someone who's hauled 100' snakes back and forth too many times, I can't say thank you enough to the engineers who made *digital snakes* possible. They can transmit lots and lots of audio channels at very high quality through a *single* standard CAT5e or CAT6 Ethernet cable. This is what's used for computer networking, and they're really cheap. Now, the systems that drive these cables are currently not all that cheap. There's a stage unit with mic inputs and preamplifiers that converts analog signals to digital data, and another unit at your mix position that decodes all the digital audio into separate analog channel outputs. Some consoles can connect directly to the CAT5 cable without a decoder box.

When our church installed one of these systems we noticed an immediate improvement in audio quality. It was also far easier to run these cables through our existing conduit.

This trend in using networking protocol to connect audio devices has greatly expanded to where you can now buy amps, processors, consoles, and so on that have Ethernet jacks built-in. You connect everything to a network and control it all from your computer. It scales upward very easily, so if you have separate areas in your church facility with audio/video setups, they can all be connected to the main audio network. It just keeps getting better and better.

They want you to plug *what* in for the service?

How often has someone walked up behind you, usually about 15 minutes before the service starts, and asked you to play back a video or audio recording? Of course you say "no problem", then start looking at the back of the board to figure it out. They usually break down into the following options:

Compact disc

CD players commonly use RCA jacks, so run two RCA > 1/4" cables into a channel or stereo return. You could use the RCA "tape" inputs if your console has them, but a channel allows you to EQ the sound and send it to your stage monitors.

iPod or other player

Use an 1/8" > 1/4" adaptor. Run it into an empty channel or a stereo/aux return. You could also use the same RCA tape inputs I just mentioned for CD players. Many consoles have USB inputs so you can connect directly.

Video camera

This depends on what audio outputs they have. RCA? Use an RCA > 1/4" adaptor into a channel or stereo return. 1/8" output? Same thing except use a 1/8" adaptor.

Computer file

If this comes on a flash drive, plug it in and try to import it into your music/media software. There are lots of different audio formats, so it depends on whether your system supports it. If the computer file was burned to a computer CDR, don't play this in your CD player unless the disc was formatted as an authentic audio CD (known as the red-book audio specification). Try it in your computer first and see if your media player can import it. Check the audio level so the volume works in the room.

Cassette

Yikes. If you still have one of these (many churches do), it will connect the same way as the CD player. Or you could just beg off and tell them it croaked a few months ago with no hope of recovery.

By the way, you can buy all these adaptors at your local electronics/entertainment store or any professional audio dealer. Don't go cheap...get the good ones. Same thing for buying cables—splurge for the good stuff as much as you can afford. It really does make a difference.

And one more thing. *Always* mute the channel before connecting and disconnecting cables.

Console connections

Someday you'll surely run into a situation where you need to check the cables on the back. When that happens, you really ought to know how things are connected, even if you're not responsible for system design and maintenance. Here is what all this looks like on the back of the board.

Main and monitor mixes

Your main house mix and stage monitor mixes have to go backstage where your amplifiers and monitor systems are. These are sent through *line returns* on the snake. Along with the XLR mic channels in the snake from stage to the console, they usually also have a few channels of line-level 1/4" cables that go back in the other direction. So if you connect the snake's line return channel 1 to

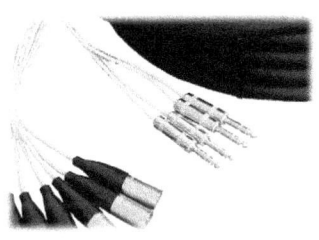

the main mix output on the console, go backstage and plug that channel 1 1/4" cable into the power amplifier that's driving your main speakers. Digital snakes work pretty much the same way, providing a number of line returns along with mic channels.

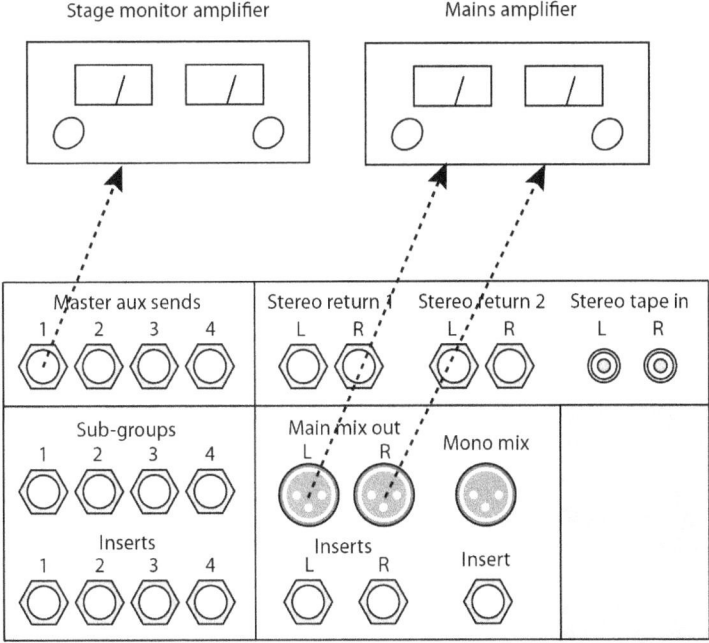

Back of console: Master section

Microphones

Unless you're running a CAT5-based snake, the individual XLR connectors on the snake plug into the channel XLR jacks on the console. To make your life easier, make sure snake channel 1 goes in console channel 1, and so on, so when your assistant plugs a mic into snake channel 14, you know it's coming in channel 14 on the console. Sometimes, though, mics need to be connected into inputs that don't match the stage numbers, so be sure to write it down and keep a chart so you don't go crazy looking for a signal.

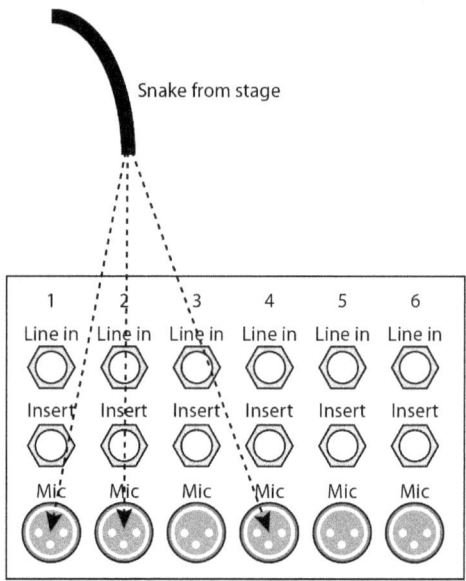

Back of console: Channels

Compressors and noise gates

These are single-channel devices; the idea is to detour the signal out to the processor, process it, then bring that signal back where it came using an insert cable. You'll notice on analog consoles there are insert jacks on each channel, subgroup, and the main mix, so you can put a single compressor on one instrument, a vocal group, or even on the entire mix. Years ago at my church we ran an 8-channel analog compressor. A few individual parts had their own compressor channel, such as acoustic and bass guitar. One compressor channel was dedicated to the pastor's mic, and yet another was plugged into a vocal subgroup.

Connect the single end into the insert jack on the console, then plug the two connectors on the other end into the device's inputs and outputs. Pay attention to the labels on these—the *send* goes to the input, the *return* connects to the output.

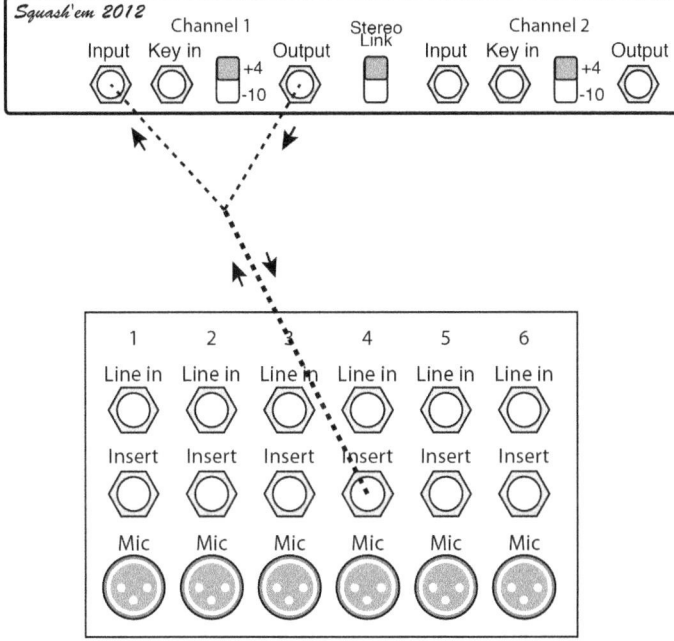

Back of console: Channels

Many compressors are two-channel units, like this one above. They can function as two independent, mono devices, or both channels can be linked into a single, stereo unit. If you insert this compressor on a stereo mix or subgroup, use the linked feature. In this case you only use one set of controls for both channels (check the manual). If you need to compress a snare drum and bass guitar, go mono and pretend they are two separate devices in your rack.

Of course digital consoles provide compressors and gates (called *dynamics*) on every channel and bus, so you don't have to physically plug anything in. Just turn on the processor.

Reverb and effects devices

These are almost always connected via aux sends and returns. Inserts are no good because that limits the processor to a single channel or group. We often add the same reverb patch to the entire vocal group, so by using the aux sends on each channel we're able to send anything we want to the reverb unit. It also allows us to customize how much reverb we get on each part simply by turning each channel's aux send up or down.

If you're using aux 1 for reverb, connect the master aux send 1 output into the left (mono) input of your effects device. Now connect the left and right outputs from the device into a stereo return (or empty channels). This gets routed straight to the mix bus. Effects devices have 1/4" and/or XLR connectors on the back. These are always balanced connections (explained earlier...did you forget already?), so make sure your 1/4" cable has the second black ring around the plug. Otherwise you'll lose signal level and open the door for outside RF interference.

One more thing...look for a switch that's labeled "+4" and "-10" or something similar. This is important because it matches input and output levels with your other gear. If it's wrong, then you'll get distortion or insufficient signal. Pro audio equipment runs at "+4", which I'm not going to explain here. What matters is to run everything at the same level, so find the switch setting that matches the best between devices.

Again, digital consoles offer built-in effects, so look through your manual to see how to route an aux to an effects module and bring that back into the mix. The concept is the same.

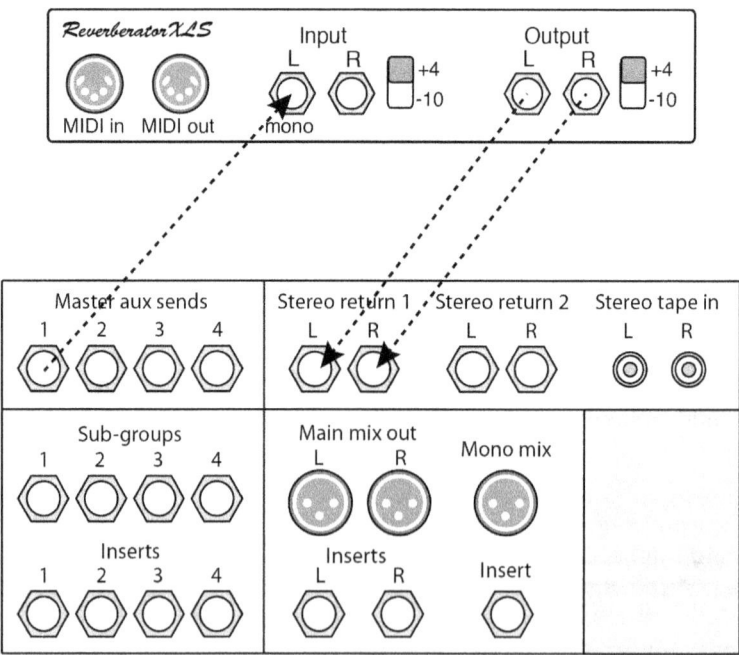

Back of console: Master section

iPods, CD players, computer audio

Most of the time these playback devices are connected to aux/stereo returns in the master section, which then routes to the main mix. They have either stereo 1/8" or two mono RCA jacks (L & R), and the console inputs are typically 1/4" or RCA. Another option is to plug them into a regular channel via 1/4" line inputs. This could come in handy, for example, if you need to route the playback into the stage monitors; using a channel allows you to turn up the monitor aux send on that channel so they can hear it.

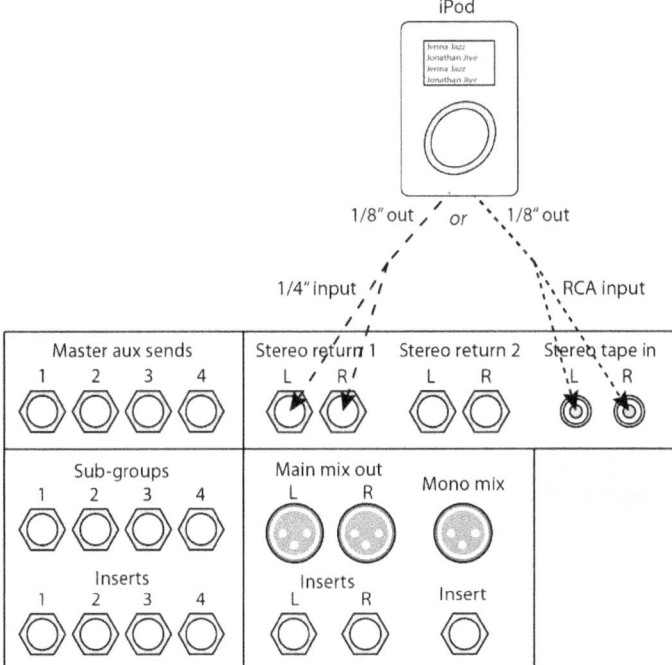

Back of console: Master section

Digital console connections

If you have a digital console and snake system, it's possible you may not use most or any of these connections. Microphone signals arrive from stage via an Ethernet cable, which plugs directly into an Ethernet jack on the snake system or console. Signal processing such as compression, gating, and reverb are built into the mixer. iPods and other similar devices will connect via USB or analog inputs as described above. The digital snake we used to have at my church ran to the console via an Ethernet cable, but then was converted to analog and connected to the digital board through multichannel connectors (due to incompatible formats between manufacturers). Our new system uses exactly one Ethernet cable from console to backstage mic preamps, then another Ethernet cable from there to our personal monitor system. There are almost no cables hanging off the back of the board—amazing. You still need to understand the concepts of how everything ties together, though. For example, if you turn on the compressor that's built-in to your digital mixer channels, it's the same thing as inserting an external compressor via the insert jack on an analog board. Adding reverb to all your vocal channels works the same, only you don't actually plug in a cable on a digital system. Remember these signal flow and connection diagrams when you're working on a digital system and it'll make things a lot clearer for you.

Power amplifiers

All you need to know about power amplifiers is that they have to crank up the signal coming from your console high enough to power your speakers. It takes a lot of juice to drive speakers, especially the big ones like subwoofers, so whoever installed your sound system should have properly matched the right amps for your speaker system and set the levels correctly. Your system may have one amplifier to run your main speakers, or there might be several amps driving various components of those speakers (low, mid, high frequencies). From time to time go take a look and make sure the red lights aren't flashing on them, though, as that means you're driving them too hard. This causes distortion (nasty-sounding sound) and is not what you want. Either pull your main console levels back with your mic trims and faders, or contact your installer to balance it out.

Amps are also needed to drive floor monitor speakers, so your aux sends from the console used for monitor mixes will be connected to these amps first, then out to the floor monitors.

These days many churches are using *powered speakers* (also called *active)* for mains and monitors, meaning the power amplifiers are built into the speaker cabinets as one unit. They're more expensive, but then you don't have to buy separate amps, and they are matched to sound great together.

Amplifiers generate a lot of heat, so they should be mounted in your amp rack with plenty of airspace around them. Look for the holes punched into the case and don't cover them up. If they're stuck in a stuffy closet they could overheat and shut down...in the middle of the service. It also shortens their lifespan. Your installer should have done this correctly, but sometimes you just never know...

Microphone Design

Although you don't really need to know much about how microphones work inside, some of this comes in handy when selecting and placing microphones. So, be brave and slog through some technical jargon.

Design types

All microphones are *transducers*, which simply means they are devices that transfer one form of energy into another. Microphones take acoustic sound waves and generate electrical signals. Other transducers in audio include tape recorders, speakers, and phono cartridges for record players.

Although the sole purpose of a microphone is to capture an acoustic event and convert it to an electrical signal, there are multiple ways to achieve this. None is particularly any better than another, and variation in mic design is deliberate so as to provide each microphone with a distinct sound. The result is an artist's palette of mics that provide different nuances to your services or recordings, so you can match just the right mic to that special soprano in your life.

Audio example 68: Comparing mics using a pair of Shure KSM32 vs AKG 414

Audio example 69: Comparing Shure SM57, Sennheiser e906, & Rode NT1000

The two primary design types are *dynamic* and *condenser*. Other variables such as materials and circuit design will also contribute to each mic's distinctive sound.

Dynamic microphones

This design employs the physics principle of electromagnetic induction. Inside the capsule of the mic a coil of wire is suspended within a magnetic field. The microphone diaphragm is attached to the coil, and when sound waves hit and move the diaphragm, the coil of wire also moves back and forth within the magnetic field. This generates an output voltage that varies according to the diaphragm movement. So, the output signal varies *analogous* to the changes in the original acoustic sound wave…yes, that's where we get the term *analog*. Amazing.

Examples of dynamic microphones include the Shure SM57, Shure SM58, AKG D112, and Sennheiser MD421.

Two variations of dynamic microphones

Moving coil

This is the most common type of dynamic mic and includes probably all of the models you see at your church.

Ribbon

Instead of a coil of wire, a thin ribbon of electrically conductive material is suspended within a magnetic field. The principle is the same as for moving coil. These are very popular for studio recording, though I wouldn't worry with them for live sound. Most ribbons are bi-directional, just so you know, meaning they pick up sound from both the front and back.

Condenser microphones

The principle of electrostatics is the basis for condenser mics, which are also known as capacitor microphones. These have two electrically charged parallel plates to transduce acoustic waves; one is movable (the diaphragm) and the other is fixed, effectively forming a capacitor. When sound waves hit the movable plate, the distance between the two plates varies. This changes the capacitance, generating a corresponding output current.

Condenser mics require a DC power source to charge the plates and power an internal preamp; this low-level amplifier reduces high internal impedance and increases the output level a bit. This is not the same as the mic preamp on the console that boosts the mic signal up to standard line level. The power source is called *phantom power (+48V)* and is usually supplied from the console through the mic cable. Look for a switch on each channel, or perhaps for a bank of channels on the back of the console. If you plug in a condenser mic and nothing happens, verify that phantom is on for that channel (mute the channel before turning phantom on/off).

Some condenser mics you might run into include the Shure SM-81, AKG 451, and the Audio Technica AT-4040.

Directionality

There are times when sound needs to be picked up from only one direction, other times from all directions. Microphones are designed to be selective in where they collect sound waves. We call these directional characteristics *polar patterns*, and there are three primary types:

- Cardioid (uni-directional)
- Bi-directional
- Omni-directional

A cardioid microphone will collect sound primarily from *on-axis*, meaning the front of the microphone. This is useful when you don't want a particular mic to pick up other sounds that are intended for other mics, or when you're running stage monitors in front of the vocalists. However, these do not completely disregard sounds from off-axis. As you get farther away from on-axis (moving toward the side and back of the mic), the microphone gradually attenuates the sound it picks up. This results not only in a signal level reduction, but also causes changes in the frequency response of the sound, altering its tone. This means that the sound will be unnatural as you get more off-axis; the term is *off-axis coloration*, and it's not something normally desired. So pay attention to where you point it. I've seen pulpits where the microphone was pointing sideways (and it sounded like it).

Bi-directional microphones pick up sound from in front and in back of the mic, but reject sounds from each side. Omnidirectional mics accept sound waves from all around, though there is a slight directional characteristic at high frequencies— meaning there is a slight roll-off of high frequencies from off-axis of the microphone.

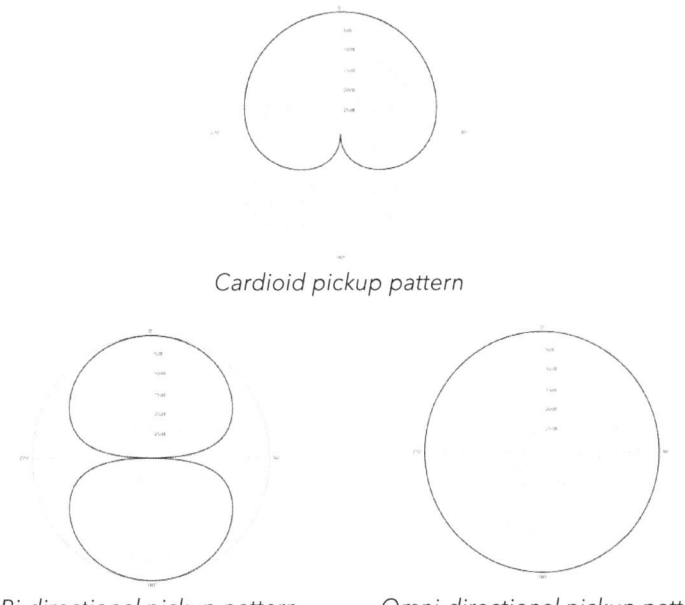

Cardioid pickup pattern

Bi-directional pickup pattern   *Omni-directional pickup pattern*

To achieve directionality, microphones have ports (openings or slots) along the outside of the casing. These ports lead to internal chambers that are designed to delay sound waves before striking the diaphragm from the inside. Cancellation is achieved when a soundwave coming from behind the microphone enters these ports while also diffracting (bending) around to the front of the mic to strike the diaphragm from the outside. When these two identical sound waves hit the diaphragm at the same time from opposite directions, the result is cancellation. Back to physics class, if two people push against each other with the same force, the result will be...zero. Nobody moves. Now, with audio signals it won't be total cancellation, depending on position relative to the mic and the fact that all frequencies have different wavelengths. This will be discussed further in the acoustics chapter, but the idea is that different frequencies will be attenuated/cancelled as the source moves around the microphone, resulting in a colored, altered sound. The lesson here? Listen carefully to each of your mics to hear what they're picking up. You might have to move a few folks around on stage to keep a mic from capturing unintended sounds from the side.

Frequency response

As audio passes through any device or system, the relative balance of low, mid, and high frequencies is affected. Ideally we usually want a device to keep the original signal intact with no change. Microphones are a different story, however. I mentioned a few moments ago that different microphones are intentionally designed to produce different sound nuances. Microphones with large diaphragms tend to reproduce low frequencies better, making them good candidates for kick drums, floor toms, and upright bass. Small diaphragm mics handle high frequencies better, so they can be ideal for cymbals. There are no rules, but these characteristics can provide some guidance when selecting microphones. Every mic you purchase comes with a frequency response chart, so check it out and compare with other mics to see how they respond differently.

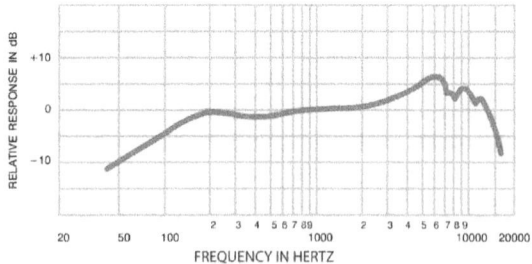

Frequency response diagram showing how level fluctuates through the audio spectrum

Something to be aware of is that when using cardioid or bi-directional microphones very close to a sound source, low frequencies are over-emphasized. This low-end boost is termed *proximity effect*. Sometimes this is cool, such as for helping radio DJs get that really deep, booming voice. For live events it just muddies your sound. There are a few ways to reduce this, such as moving the mic away a bit or by switching in a *low-cut filter*. Many microphones have a switch that attenuates low frequencies, usually around 75Hz or 80Hz, and your console will also have these on each channel. Digital consoles allow you to dial in the frequency you want to start attenuating, so it could be set higher for a female vocal and lower for an acoustic guitar or piano.

Transient response

How fast and accurately does the diaphragm respond to acoustic waves hitting it? It takes time for the diaphragm to move and for the electrical signal to be generated. Audio transients are very brief bursts of signal found in all sounds, especially percussive events such as drums, guitar strums, even piano chords. They are often so short many microphones simply ignore them. High frequencies have very short wavelengths and fluctuate very quickly. The quicker the response of the microphone, the more accurate the reproduction. Condenser microphones exhibit a quicker, more accurate response that reproduces clear high frequencies, whereas dynamic mics are more sluggish and tend to round off the waveform, resulting in a smoother, more mellow sound. Neither of these is better than the other—they're just different. So, if you place a condenser mic on a trumpet and it peels your ears back, replace it with a dynamic and enjoy the mellower sound. If you're looking for a crisp, bright sound from your cymbals, put a condenser on them.

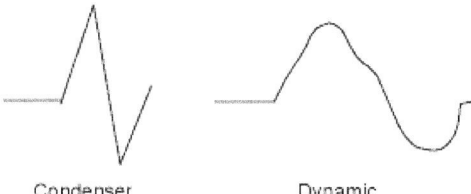

Condenser   Dynamic

Transient response differences between condenser and dynamic mics

Audio example 70: Comparing dynamic vs condenser mics

Overload

Just like with any audio device, there is a limit of how much signal level a microphone can handle. A mic is overloaded when the SPL level (sound pressure level) is so high it distorts (clips) the diaphragm and/or electronics. High level sources such as drums or brass can potentially cause problems, though these days most mics are pretty robust and not often an issue for live sound. Dynamic mics feature a higher overload tolerance than do condensers. If you're distorting the mic itself (as opposed to overloading the mic input on the console), either move the mic farther away or use an attenuation pad switch if the mic has one. I once recorded a vocalist that had such a powerful voice she was overloading the mic—and it was a very nice, very expensive condenser. The mic pad saved the day and we got a great sound from this terrific singer.

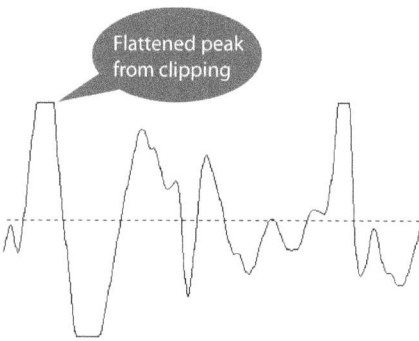

The waveform gets clipped when the device can't handle the high signal level. This actually changes the frequency components in the sound, resulting in a different tone we call distortion.

Impedance

This little-understood term is important in the design and connections for audio equipment. For now we'll just leave it as the mic's ability to provide a certain signal "strength" as compared to what the mixer wants to see. Another way of saying this is that the console input is expecting a certain signal strength, so the microphone output must be designed to match this for optimum transfer of signal. Remember when you go to a concert and they have restricted gates and entrances to control the crowd going in? Think about if they either closed these nearly shut or opened everything completely. People would pile up trying to get in or you'd get a stampede. You get the idea. Maybe.

In more practical terms, all professional microphones are low impedance (low-Z), so don't use high-Z mics, which usually have 1/4" connectors and act as radio antennas for any available broadcast that happens to be floating through the room. Low-Z mics are much better at preventing outside interference and extraneous noise (motors, fluorescent lights, radios). High-Z mics also suffer from high-frequency loss over distance.

Balanced microphone cables

Professional low-impedance (low-Z) microphones use mic cables employing two signal-carrying wires in addition to a ground wire (shield). These signal wires are twisted around each other throughout the cable, and the shield is most often braided around the two wires. This provides maximum protection from outside noise interference, or RF (radio frequency).

How does it do this? Audio signals are AC current, meaning they alternate positive/negative between the two signal wires. Any outside interference leaks into the cable as a common polarity DC signal. When it arrives at the end it's cancelled out because all balanced audio gear (professional equipment) is designed to accept AC signals only. The shield in the cable drains extraneous noise by shunting it to ground.

Balanced audio requires a connector with three points, so we use either XLR connectors or TRS 1/4" (tip, ring, sleeve), and the three pins are numbered so as to match on each end of the cable. Pin 1 is always the shield (ground); pins 2 and 3 carry the alternating polarity audio signal.

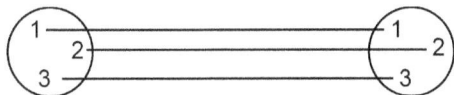

Each XLR connector has three pins for the three wires in a mic cable

Phantom power

As we described a bit earlier, condenser microphones require an external power source, usually 48 volts DC. This is sent from the console through the mic cable to the microphone, but does not damage the mic (or anything else it's plugged into). Audio equipment works with alternating current and looks for the *difference* between the two wires; phantom power is DC. Since there is no difference at the input it is ignored by the device, meaning it won't damage anything plugged into that channel.

So it's not a huge deal to leave phantom power turned on, but if the console has individual channel phantom switches I leave them off unless a condenser is plugged in.

PZM (Pressure Zone Microphone)

Ever notice those strange flat mics? These are called PZM mics, though the term PZM is actually a product trademark of Crown, Inc. The generic term is *boundary microphone*, and refers to a microphone where the mic element (diaphragm assembly) is mounted on a flat plate. The concept is to reduce phase cancellations that occur when using a traditional microphone stand. With a stand, the mic is elevated above floor level by several feet. The sound reaches the mic, but also bounces off the floor and into the mic somewhat later than the original wave. Back to acoustics, when two identical waveforms arrive at different times, phase cancellation occurs, which means the tone of the sound is altered in a negative way. With a boundary mic, there is no phase-altering reflection from the surface since the mic is directly mounted on that surface.

PZMs are popular for miking underneath a grand piano lid. They can also be useful for stage productions, such as the annual kids' Christmas Pageant, by placing them on the floor in front of them. (On second thought, they'd probably just stomp on them and enjoy the cool sound it'll make through the speakers.) PZMs do very well picking up everything that happens on stage. Just be sure to put a piece of foam or rubber underneath to help reduce vibrations from movements on stage as your actors run around crazy-like.

I installed a system in a small church years ago where they used an upright piano. The piano was boxed in on three sides by a 4' high knee wall, so I taped a PZM to the wall facing into the back of the piano. It worked okay...not many options there.

Microphone switches

Pay attention to the various switches on your mics and make sure they're set to what you want. Polar patterns are typically set to cardioid (front pickup only), and filters and attenuation switches are off until needed. Not all microphones have these switches, so if you don't see a polar pattern switch on your SM57, don't tear it apart looking for one.

- Polar pattern select (some condensers allow you to change the directional pickup pattern)

- Attenuation pad to reduce incoming signal level (condensers)

- Low-cut filter to attenuate low frequency sounds (condensers & dynamics)

Direct box

These small boxes allow you to connect electronic instruments such as keyboards or guitars to the mixer. You can buy *passive* or *active* models. An active DI requires phantom or battery power and uses active electronics to provide a "hotter", typically

higher quality signal output. You can find very nice passive models as well; the point is to buy good ones—the cheapo models sound terrible. A decent DI can be had for under $100, such as from Radial or Nady.

Direct boxes have a 1/4" input jack, where you plug in the guitar, and an XLR output jack where you connect a standard mic cable to the stage snake. Once you plug it all in and turn on the channel, listen for a hum (ground loop). If you hear it, flip the ground lift switch on the box. It doesn't matter which setting you use, just pick the position that reduces the noise.

Some DIs also have an instrument/amp switch. If you're plugging in a guitar or other instrument, set this to "instrument". You can, however, take the output of a guitar or keyboard amplifier, which is a much higher signal level, and run that through the box. In this case set the switch to "amp".

And in case you're wondering, "DI" stands for "direct injection", meaning the box takes your signal directly into the system rather than through a microphone. Nobody says "direct injection", though, so I wouldn't try it out on your friends. They'll just stare at you.

Microphone technique

So, now that you're impressed with all the various features and specifications of microphones, you're back to the basic question: How do I use these things to get a great sound? Microphone technique is very subjective; there are a few guidelines, but much is up to you, your team, and your particular situation. The rule is not to worry about rules too much and just experiment. Sure, I'll give you a couple of things to keep in mind that might cause problems with your sounds, but generally you should just try lots of things to see what you like.

The first thing I'll tell you is to go listen to the instrument before you pick up a microphone. What does it *really* sound like? Walk around, move your head, ask the musician where the sweet spots are as all instruments have different radiating characteristics (where the sound comes from). Once you find a spot you like, go pick a mic that best matches that sound. Not sure? No problem, just try a few. Don't immediately begin twirling EQ settings on the console. Listen first, find the right mic, then find the right place for that mic.

For live events, such as a church service, you really don't need to be as picky as we are in a recording studio. This doesn't mean that anything goes, however, so take these guidelines and ideas and apply them to the miking suggestions I provided earlier in the book. Little things can help a great deal.

Placement options

In terms of where you put the microphone, the main variable is the distance from the source to the mic.

- Close miking
- Distant miking
- Ambient miking

Close miking

You'll be close-miking nearly everything most of the time for church services, with a few exceptions such as a choir or large ensemble. I'll explain the pros and cons for distant miking in a bit, but for now we'll concentrate on keeping mics close to each source. This usually means anything from a couple inches to a few feet away.

A microphone placed close to a source gets a clear, direct sound without lots of surrounding noises. As you move the mic farther away you'll begin hearing more of these sounds, such as other instruments and singers on stage. Close placement also means that the mic won't capture much of the room's acoustics, meaning reflections and reverberation. This is a good thing for live events, where the room will affect the final mix anyhow. Keep it clean up to that point.

Audio example 71: Close vs distant miking

Experiment with slightly different distances from the source. For example, a kick drum sounds very different when mic'd near the beater head vs farther back in the shell. Place a mic on a mounted tom about an inch away, move it back to three inches, and listen to the difference. Take some time to play with it; it's far better to get the sound you want from the mic rather than trying to EQ it to death.

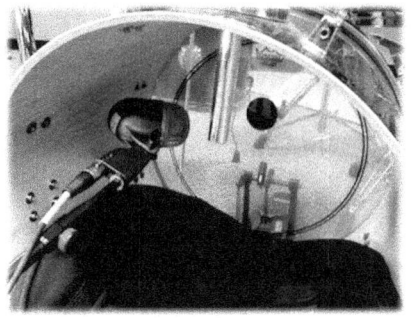

Audio example 72: Kick drum mic close to beater, then farther back in drum

How close is too close?

Placing a mic too close, though, results in a colored tonal quality—it doesn't sound natural. A good example is the piano, which is large, complex, and needs space and time for everything to radiate and merge. So, while we could place a mic 6" directly over one section of the strings, you're not going to get the entire frequency range of the instrument; the result is an uneven piano sound. Miking very close also results in a low-frequency boost called *proximity effect,* which makes it sound boomy and muddy.

Solutions

- If it sounds too "in your face", back the mic off a tad or angle slightly to one side.

- Boomy sound? Pull the mic away a bit or turn on the low-cut filter.

- Muddy? Try some EQ, reducing the low-mid range where this usually occurs (see the section on EQ).

- Doesn't sound full or natural? Try backing the mic away and experiment with placement. Small changes can make a big difference.

Audio example 73: Proximity effect (bad, then cleaned up)

Audio example 74: Tom too close, then a bit farther away, then too close

---

Leakage

Leakage is when a microphone picks up sound from something other than its intended source. We pay a great deal of attention to this in the recording studio, but it can also cause problems for live events. Leakage into a microphone colors both the leaked-into mic as well as the mic focused on the other source. In other words, if you've got several mics on stands for singers, and they're standing a few feet away from their mic, you'll get a mash of everything into all the mics. Or perhaps the brass section is being picked up by the choir mics. Or the choir mics are picking up the piano which is sitting right in front of them. In this case, you turn up the choir mics and hear more piano (and not a very good sounding piano at that, destroying the main piano mic sound). Your goal is to have control over the level and tone of each instrument or part, so reducing leakage is a major objective.

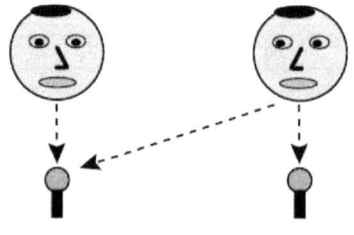

Distance can be your friend here. The farther away a sound is from a mic, the less it will be picked up. Every time you double the distance between the mic and a sound source, the signal level drops around 4–6dB. So if you place a mic about 6 inches from an instrument, and other instruments are several yards away, you can do the math and see how effective this can be. It's not perfect, though. If multiple sources are making noise in the same room, you're gonna hear them in all your mics, especially condensers.

Solutions

- Keep sources close to their intended mics.

- Try to spread sources out on stage if you can.

- Point the mics away from other sources on stage.

- Block loud sounds with plexiglas panels, which we often do for drums.

- Make sure you don't have any bi-directional or omni-directional microphones (though these are fairly rare). Stick to cardioids that pick up from one direction only.

Audio example 75: Leakage into a mic channel

Phasing

Yikes, this is a tough one. Phasing is when two identical or similar signals combine and destructively affect each other. This happens whenever you have a time delay between the two signals. For example, with two mics picking up the same sound source, any difference in path length between the source and each mic causes a timing discrepancy between the two signals, therefore inducing phase shift.

In plain English, this means that if you have one mic farther away than another mic, it takes longer for the sound to get to the second mic; when both signals get mixed at the console they don't play well together because they're not quite the same due to the timing difference. Since frequencies have different wavelengths the phase shift varies throughout the audible spectrum. The result is a succession of

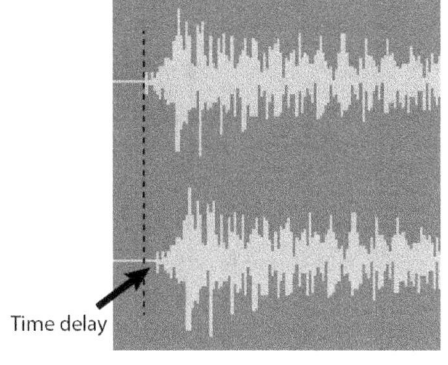

Time delay

dips and peaks in the frequency response (constructive and destructive interference). We call this a *comb-filter effect*. It sounds very colored in tonality, but it usually takes practice for people to recognize it.

If you remember sine waves from school (sorry about that), think of two identical sine waves. Now slide one of them over slightly—the peaks and troughs don't quite line up anymore. When you add them algebraically (another apology here), you don't have a sine wave anymore. This will sound different from the original, usually in a way that you won't like.

Phasing also occurs when your mic and/or source is close to a nearby surface. Say you're miking a guitar and it's located near a knee wall on stage. The guitar sound goes directly into the mic, but it also strikes the nearby wall surface and then reflects into the mic. Since it takes longer for the reflection to arrive, you've got two copies of the guitar signal that are out of time with each other. The microphone blends these together and sends the result to the console, where you go "yuck...what happened?". Pulpits and music stands cause the same problem, where the sound splashes against the hard surface and bounces back into the microphone. Cover it with a thick cloth or try to change the angle so it directs the reflections farther out of line.

Earlier in the book we discussed miking a drum set. Lots of mics in a small area, so you'll get phasing and leakage big time. You can't solve it all, but turn one mic on, then another, and then another until you hear a weird tonal change. This is where you probably have some phasing issues. For example, listen to the snare mic alone, then add others into the mix. At some point your snare sound will change, meaning either you've got leakage (you now also hear the snare from the overhead or tom mic) or phasing (from multiple mics picking up the snare at different distances). Some of this is unavoidable and normal, and it's tough to learn the difference. Keep mics close to their targets, face them away from other stuff as much as possible, try moving things around a bit, and then send an email to your church's prayer chain.

> Audio example 76: Phase delay issue with overhead mics on a drumset

Solutions

- Follow the 3:1 Rule: For every one unit of distance between the closest mic and its source, the distance between that mic and any others needs to be at least 3 times that mic-to-source distance. So, make sure any other mics that are placed near a source are located farther away from the first mic.

- Use one microphone and be careful not to locate it close to a large reflective surface. Sound bouncing back into the mic from a nearby wall causes phasing the same as using two mics.

- Consider a boundary mic (PZM) for certain situations as these are designed to minimize phasing issues. However, you're not going to mic your singers and acoustic guitar with these.

- Look for nearby surfaces that can cause delayed reflections. Put something fuzzy there or move away from the surface. We discussed this early on in the miking setup section.

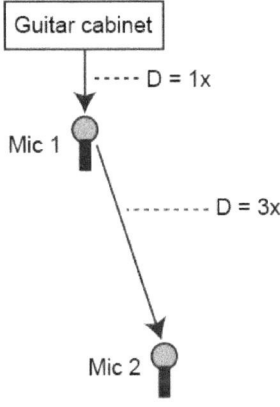

3:1 rule for using more than one mic on a source

Audio example 77: Listen how the snare sounds different when OH mics are on.

Sibilance

When close miking vocals, be careful of side effects such as sibilance and explosive pops. Sibilance is excessive energy from the "s" sound, generally around 6-7 kHz. Pops come from hard consonants such as "p" and "b" where a great deal of wind is expelled downward from the mouth. Have your vocalists hold the mic in front of the chin, not in front of the mouth where this air blows out. This helps for both of these conditions. If you're still having low-end pops and booms, switch on the low-cut filter on the console channel. For sibilance, you can reduce the 6-7kHz region with an EQ, but on many consoles this tends to yank out too much of the sound because they can't dial in narrowly enough. Digital EQs offer a "Q" control for each EQ band, which is how you can narrow the region you're changing. If you happen to be running a console with software plugins, insert a de-esser to take care of these. Dynamic mics also reduce sibilance issues as they don't have the high frequency response that condensers have.

Audio example 78: Sibilance

Audio example 79: Removing a vocal pop using a low-cut filter

Distant miking

For distant miking the idea is to get a more rounded sound from your source blended with some room ambiance (reverberation). The microphone could be anywhere from a few feet to nearly across the room. Of course, for a church service it's not very practical to do this, as opposed to a recording session where you may only be working with one instrument at a time. You'll mainly use close-miking, but some of this can come in handy from time to time, especially if you're doing some recording with your team.

Advantages

Distant miking picks up more overall sound of the source, whether it is a single instrument or an entire ensemble. A sound source, especially more complex ones such as the piano or a choir, needs space for the radiating sound to evolve and blend naturally, thus creating their characteristic sound we're accustomed to hearing.

It also includes the surrounding acoustical environment which blends with the direct signal from the source. This balance is determined by the size of the source as well as size and characteristics of the hall or environment. Hear the difference between singing in a bathroom and singing in your sanctuary?

However, if the room sounds bad, you've got a couple of options:

- Treat walls with absorptive material to reduce acoustical reverberation.
- Move the mic closer to the source.

Audio example 80: Miking a choir from several feet away

Boundary interference

By now you know that sounds radiating from a source not only travel directly to the mic, but also to surrounding surfaces such as walls, floor, and ceiling. These reflected waves will cause phase cancellation and attenuation at certain frequencies. Although this occurs when close-miking, it's also an issue when the mic is farther away.

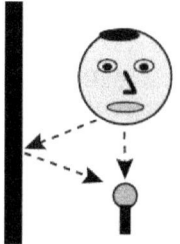

Solutions

- Install absorptive material on your walls. Sound panels made from compressed fiberglas and covered in fabric are fairly inexpensive and help reduce excessive reflections. I'll talk about this in the acoustics section, but note that high and low frequencies are different animals and require different materials and strategies for treatment.
- Move the mic closer to the source—this will reduce the room effect on the sound.

Ambient miking

The concept here is to capture the sound of the room in order to add its natural acoustics to your mix. Mics are placed farther away from the source to get more of the environment than the direct sound. Usually a pair of condenser mics are hung from the ceiling, or you can tape boundary mics to the side or rear walls (just make sure the stage manager or facilities Deacon is off-duty or they'll freak out). For churches, this is most often done to capture congregational singing to blend into the separate mixes feeding your building distribution, broadcast, or web streaming. Since those people are not in the sanctuary, this helps them get a more complete experience of the service. We might also feed ambient mics into the personal monitor mixers so they can hear the room, including congregational singing, and not feel so isolated.

Stereo microphone techniques

Many churches run their sanctuary mix in mono, not stereo. Even so, there are times when a stereo mic setup works better than a single microphone. It's especially recommended in situations where you want to use two or more mics, but need them to work together in a cohesive way (meaning mono-compatible). A few examples include a drumset, small ensemble, or large choir. You can't just throw two microphones at a source without encountering problems, so let's go through some concepts of stereo and how to set up your mics accordingly.

Directional cues

Humans can tell where a sound is coming from by two clues:

Intensity differences

Sound (primarily mid to upper frequencies) arriving from the right reaches the right ear at some intensity level. It must travel farther to reach the left ear, during which time the intensity drops a bit. This is due to the distance between the ears as well as physical blockage by the head. Also, a large portion of the sound that reaches the opposite ear is from room boundary reflections, in which case the intensity level drops significantly. In other words, it's louder in the right ear, and so our brain decodes this as a right-side sound source.

Time of arrival differences

Sound arriving from the right side reaches the right ear earlier than the left ear due to the spacing of the ears. It takes a bit of time to travel to the other side, and so the brain also decodes this as a sound coming from our right.

We use these two concepts to configure microphone setups that capture directional cues. Even if you're not running stereo, they are effective for miking sources that need more than one microphone, the goal being to reduce phase issues. I'm only going to include a few techniques here that you would most likely use in church; there are other variations best suited for recording situations, but that's not our main focus. These techniques were developed specifically to maintain accurate left-right imaging while reducing problems such as phasing, so try to follow them as closely as possible. Even though there are nuances in how each of these sounds, it doesn't make much of a difference for a live situation. Try one, try another, and see if it matters to you.

X-Y

X-Y is a *coincident* technique, which means two microphone capsules are arranged directly over top of each other. Use two identical condenser microphones with cardioid polar patterns angled left and right 90° (or larger); the capsules (not the body of the mics) must be vertically over top of each other. If not, the slight distance between them will cause phasing issues.
Since there is no distance between the two mic capsules, directional cues come from intensity differences only. You can use two mics mounted together, as in the photo here, or one stereo microphone where the two capsules are built into the housing. This technique is the best for ensuring mono-compatibility; if your church system is mono, these two mics must capture and combine everything without phasing issues. I use XY for my drum overheads, giving me a rock-solid mono blend of the cymbals and overall kit.

ORTF

This common type of *near-coincident* setup features two condenser, cardioid mics with capsules placed 17 centimeters apart (about 6.7 inches) with an angle of 110°. This spacing corresponds to the distance between our ears (in case you don't have a ruler handy). Low frequency information comes largely from intensity differences, though high frequencies 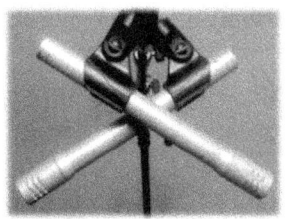 are determined through time-of-arrival cues since this frequency range is highly directional. It sounds more spacious than XY, but it's not the best for mono compatibility. Turn one of these on, pan it center, and send to the mix bus. Now add the second mic, also panned center. If it sounds thinner or otherwise not good, change the setup to an XY.

Spaced pair

When you place two microphones several feet from each other, both facing directly toward an ensemble, you've got a *spaced pair* configuration. This technique primarily follows the time of arrival principle, though intensity differences also contribute to directional cues. It seems easy enough, but remember the 3:1 rule—keep mics away from each other roughly  three times the distance between one mic and the source. So, if you set a mic four feet in front of your choir or orchestra, move the second mic away twelve feet from the first if possible. You may have to compromise on this, but do the best you can. You certainly don't want two mics about five feet in front and five feet from each other or you'll get phasing issues. For really wide sources you might run three or more mics (see photo). As with all ensemble miking, use condenser microphones only; dynamics don't have the range to pick up the source from several feet away.

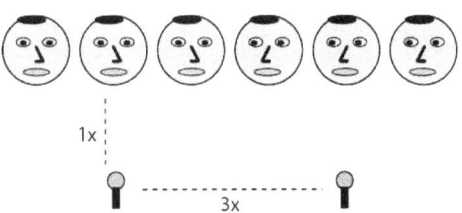

EQ

Of all the various signal processors you'll use when mixing Sunday mornings, such as compressors, gates, and effects, equalizers are the fundamental tool you need to learn. Get a solid grasp on how to dial in the tone for your vocals and guitars, get them balanced nicely, and you've nailed most of a good mix. Let's dig deeper into what makes them tick.

An equalizer, or "EQ", is a circuit that changes the frequency response of a sound by boosting or cutting selected frequency bands. Think of it as a sophisticated tone control that allows you to make a sound brighter, less boomy, and so on. To understand how it works, you first should get the concept of what actually makes up a sound.

There is a wide range of frequencies that are audible to humans—anything vibrating between 20Hz and 20kHz falls within our hearing range. All sounds coming from musical instruments, voices, or traffic on the highway produce a number of frequencies which fall somewhere within that band. What makes a trumpet sound different from a fog horn is the difference in *which* specific frequencies and *how much* of each frequency is included in that sound. In other words, the harmonic content of that sound is what makes it unique. We call this the *timbre* of a sound.

With an EQ, we can actually raise or lower these frequencies within a sound—this alters the harmonic structure and therefore makes it sound different. You're not turning a flute into a guitar, but you can make that flute brighter by boosting the upper frequencies in that sound.

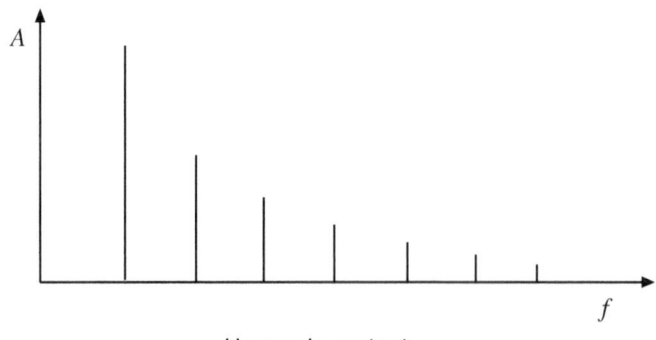

Harmonic content

A very simplified graph showing individual frequency components (sine waves); low frequencies are louder while the higher components get progressively lower in level. Real sounds are made of lots of these components in a very complex pattern.

EQ   145

Equalizer types

There are a few main EQ designs that give you control over a signal's harmonic structure, and each of these employs different types of *filters*, which are the circuits that actually do the work. First, let's introduce the EQs you'll run into, then I'll explain the filters they use.

Bass & treble

The simplest EQ around, this type gives you control over the treble (high frequencies) and the bass (low frequencies). That's it. This is the most common type found on home and car audio systems. They use *shelving* filters (explained a bit later). You won't find these in the studio or on your console as they don't provide enough control.

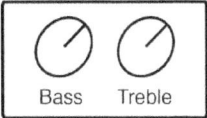

Graphic

Graphic EQs are easily identified by the row of vertical sliders that are used to boost and attenuate specific frequencies. These sliders are preset at certain intervals (frequencies) throughout the frequency spectrum, so you can only select what's there to change its level—you can't change the exact frequencies. These are found in some home audio systems, car stereos, as well as professional applications. They're popular for "tuning a room" when an engineer adjusts the speaker system to compensate for acoustic issues in a particular room or hall. You determine which frequencies are causing problems, such as persistent feedback, then reach for that particular slider to turn it down. Most digital consoles provide graphic EQs on buses (aux sends, groups, main mix).

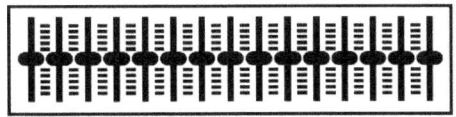

Graphic EQ

Parametric

Parametric EQs are the most common type found on mixing consoles. There are parametric EQs for each channel, though with digital consoles you'll typically have them available on all channels, sub-group buses, aux buses, and the main mix. This is the most complex type of EQ, usually employing both *peaking* and *shelving* filters, and provides not only the option to boost or cut certain frequency ranges, but also to dial in the exact frequency region you want to work on. Whereas graphic EQs come preset for certain bands, parametrics allow you to move around and find where you want to work.

Parametric EQs divide the audio spectrum into a number of bands (regions). You always have high and low frequency controls; sometimes these are fixed at a specific frequency, other times they are adjustable. At least one mid-range band will provide boost/cut along with frequency select, allowing you to fine-tune exactly which frequency area to work with. Often you'll get a third control, *bandwidth*, that determines how wide an area to affect.

On the diagram below, look at the largest section labeled *low-mid freq* and you'll see these three controls. On consoles, usually the mid-range EQ uses peaking filters while the high/low controls are shelving. Sometimes this is switchable between shelving and peaking—check the console or the manual.

Parametrics give you very precise control over your sound, but it takes some understanding and practice to use them well. Run some signals through a channel and play with the various controls. Keep reading through this section and I'll show you more of how they work and what they sound like.

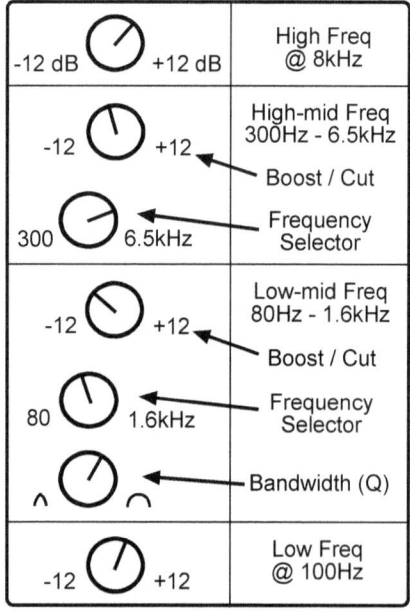

Parametric EQ Module

EQ   147

Filters

Peaking filters

Peaking filters allow you to boost or attenuate a range of frequency components centered around a specific point, called the *center frequency*. You can't operate on just one frequency, though—it always affects a certain number of frequencies on either side. This creates a bell-shaped curve and is known as *bandwidth*. The bandwidth of frequencies affected is called "Q", for *quality factor*, and it can be widened or narrowed in many parametric EQs.

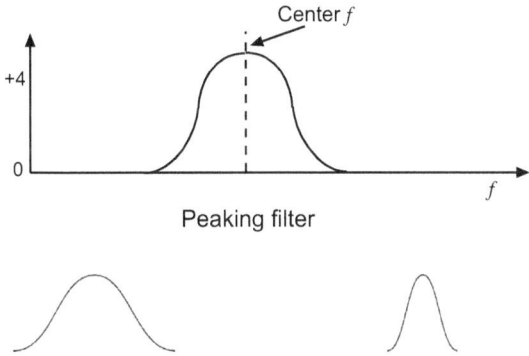

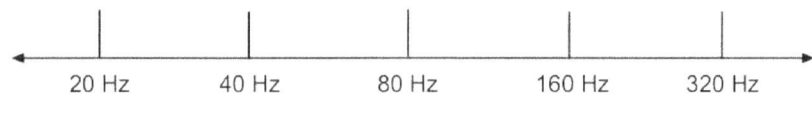

The wider bandwidth on the left will affect a larger portion of the sound, while the much narrower bandwidth on the right targets a specific, small region of the sound.

The term Q is easier to use instead of precise frequency counts, because the number of frequency components per musical octave doubles as you go up the scale. So, if you boost an octave in the low end, you might be adjusting forty frequency components, whereas in the high end you could be working on eight thousand frequencies.

20 Hz	40 Hz	80 Hz	160 Hz	320 Hz

Frequency doubles with each octave

They both sound like musical octaves to us, so we need a system that compensates for the difference in actual frequency components. Q does this quite nicely by providing a simple number to refer to. Most consoles will label this control either with the Q value or with a graphic that looks like a bell-curve.

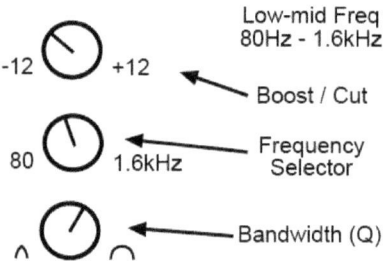

All this means is that you can dial into any specific frequency location you want to fix or enhance. It might be a small slice, such as reducing a 60Hz hum, or it might be a wide chunk of your sound, such as boosting high frequency sparkle on the piano.

Shelving filters

Shelving filters are often used for high and low frequency controls. They allow a boost or cut at a selected frequency (*turnover frequency*) at which point the boost or cut remains constant throughout the rest of the spectrum. So, while a peaking filter will affect a range of frequencies and taper off on either side of your selected band, the shelving filter affects everything equally from the turnover point and beyond.

On the graphic below, the vertical axis represents amplitude (signal level) and the horizontal value is frequency range from low to high. Everything to the right of the turnover point is what's being boosted. Everything to the left of this remains at 0 (unity), meaning the original level passes through unchanged.

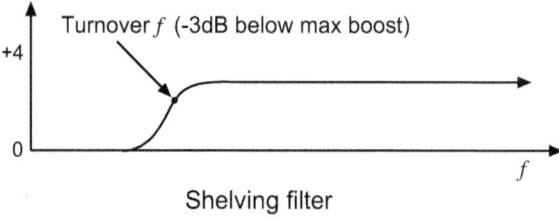

Shelving filter

EQ   149

Notch filters

Notch filters are peaking filters with an extremely narrow Q. They are designed to cut deep into a sound to eliminate a specific, narrow frequency range. One common application is getting rid of a 60Hz hum (go ahead, hum at 60Hz...that's what it sounds like). I've also used these to notch out a ringing overtone from cymbals and even guitar amps (yeah, that was a strange one). The idea is to surgically remove an offending sound without interfering with the music if it can be helped.

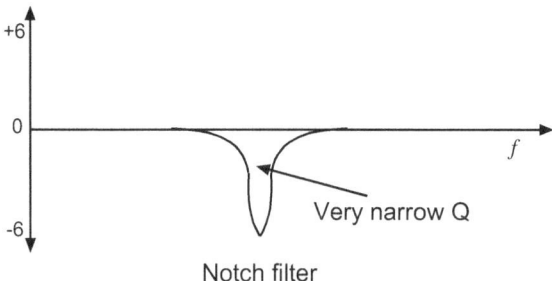

Notch filter

High- & low-pass filters

These work differently from the others in that they provide no boost function. All they do is sharply attenuate (reduce) frequencies above or below a certain point in the frequency spectrum. A high-pass filter set at 100Hz (the *cut-off frequency*) will attenuate all frequencies below 100Hz, and a low-pass filter set at 8kHz will attenuate everything above this point. The rate of attenuation beyond the cut-off frequency is called the *slope*, which is set at 6dB/oct, 12dB/oct, 18dB/oct, or 24dB/oct. Thus for every octave beyond the cut-off frequency the signal drops 6, 12, 18, or 24dB; higher numbers result in a quicker attenuation. There is no way to design a perfect "brickwall filter" that magically slices everything at the cut-off frequency. Actually, the cut-off frequency is about 3dB down from unity gain, meaning the attenuation begins slightly before this frequency.

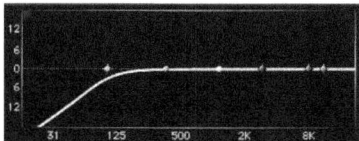 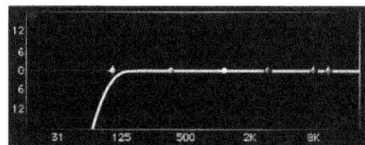

Gentler slope at 6dB/oct on the left compared to 24dB/oct on right

Low-cut filters (also called *high-pass filters*) are great for getting rid of room rumble, low-end leakage from other sounds on stage, ground loop hum, and that annoying pop from a vocal getting too close to the mic. All consoles provide a low-cut filter on every channel, with the cut-off frequency usually preset somewhere around 80Hz on analog boards. Most of your sound sources do not have anything below 100Hz, so you won't miss anything by getting rid of it. Take a look at my church console and you'll see the low-cut filters switched on for every channel except kick drum and bass; the result is a cleaner sound without all that low-end noise coming through the mix.

Cut-off frequency (-3dB)

0
-3
-6
-9
-12

f

High-pass (low-cut) filter

Let's hear what they sound like

There are lots of EQ demonstrations in the audio examples I've provided. Here, I'll show you what the EQ setting looks like while you hear the result. These are pretty extreme settings in order to make them easier to hear; if you find yourself boosting or cutting up to 9dB, then start over and work on your stage and mic sound.

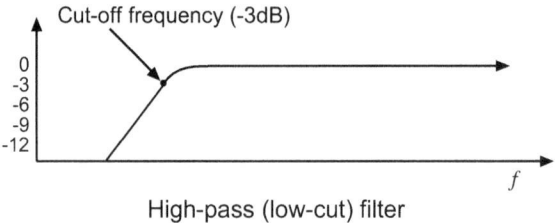

Audio example 81: Flat EQ setting (no change)

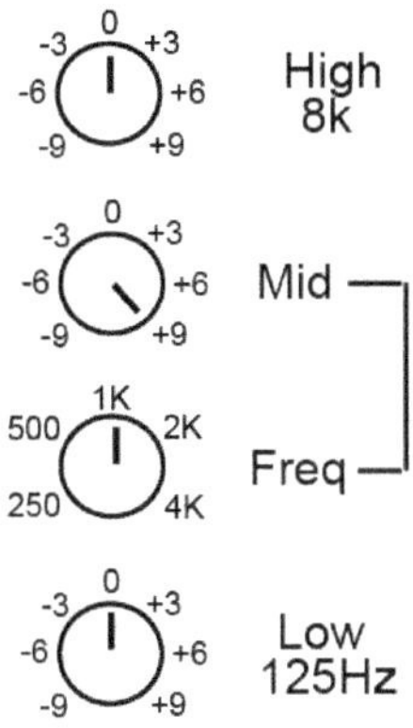

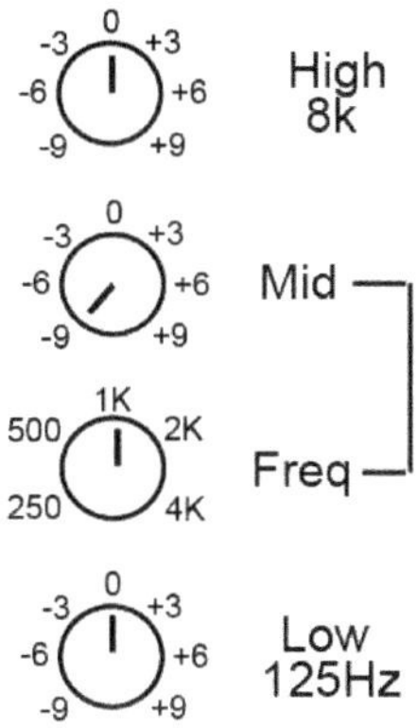

EQ   151

High
8k

Mid

Freq

Low
125Hz

High
8k

Mid

Freq

Low
125Hz

Audio ex 82: 9dB boost at 1kHz   Audio ex 83: 9dB attenuation at 1kHz

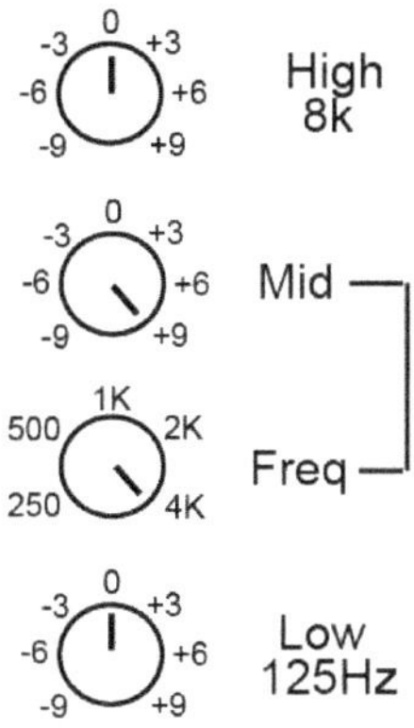

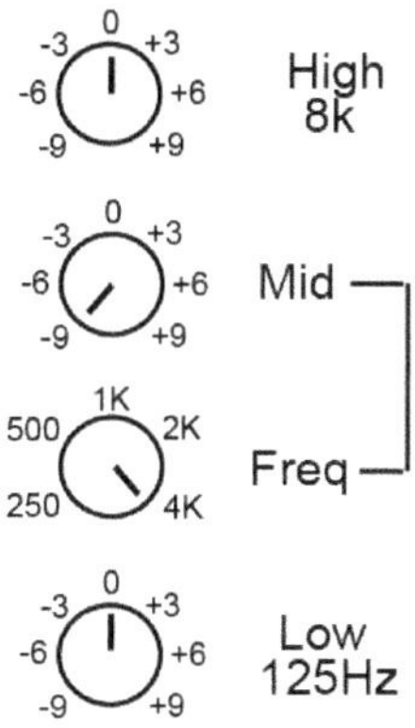

High
8k

Mid

Freq

Low
125Hz

High
8k

Mid

Freq

Low
125Hz

Audio ex 84: 9dB boost at 4kHz   Audio ex 85: 9dB attenuation at 4kHz

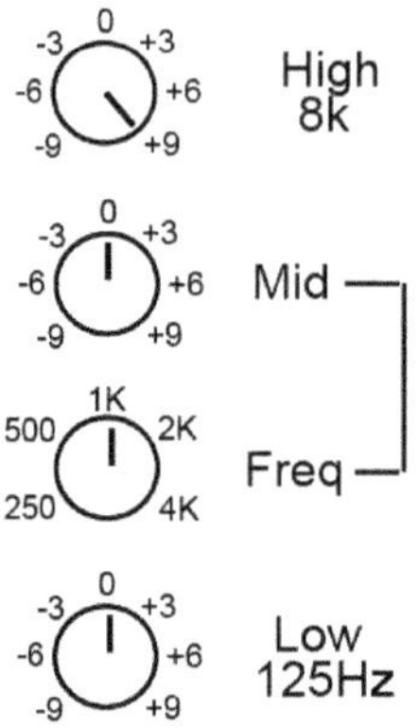

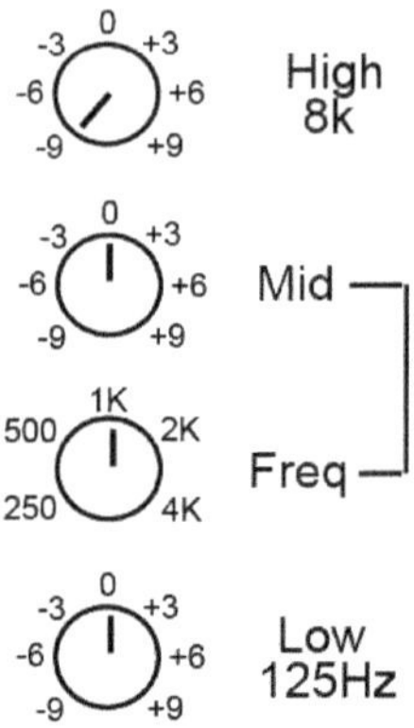

-3 0 +3
-6 +6
-9 +9
High
8k

-3 0 +3
-6 +6
-9 +9
Mid

500 1K 2K
250 4K
Freq

-3 0 +3
-6 +6
-9 +9
Low
125Hz

-3 0 +3
-6 +6
-9 +9
High
8k

-3 0 +3
-6 +6
-9 +9
Mid

500 1K 2K
250 4K
Freq

-3 0 +3
-6 +6
-9 +9
Low
125Hz

Audio ex 86: 9dB boost at 8kHz	Audio ex 87: 9dB attenuation at 8kHz

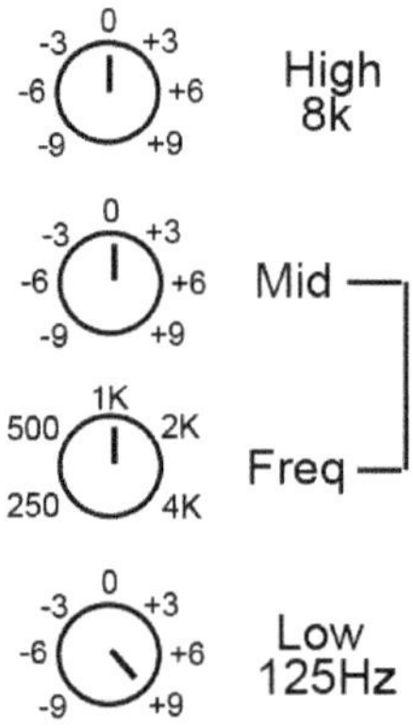

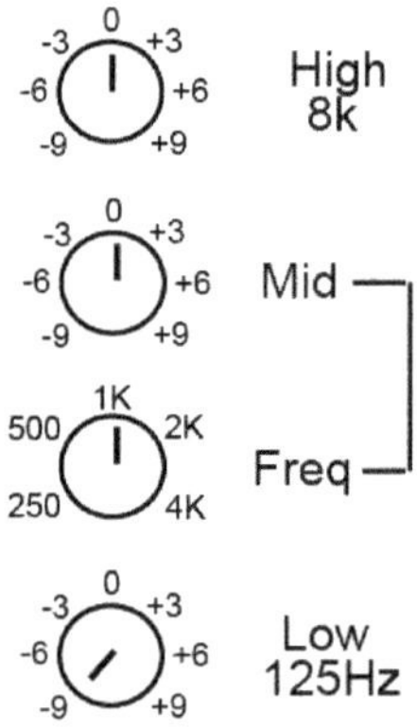

-3 0 +3
-6 +6
-9 +9
High
8k

-3 0 +3
-6 +6
-9 +9
Mid

500 1K 2K
250 4K
Freq

-3 0 +3
-6 +6
-9 +9
Low
125Hz

-3 0 +3
-6 +6
-9 +9
High
8k

-3 0 +3
-6 +6
-9 +9
Mid

500 1K 2K
250 4K
Freq

-3 0 +3
-6 +6
-9 +9
Low
125Hz

Audio ex 88: 9dB boost @ 125Hz	Audio ex 89: 9dB atten @ 125Hz

EQ 153

If your console provides bandwidth control, you can choose to adjust a big chunk of the sound or just a small slice. These three examples include a file with no EQ, EQ cut with a pretty wide bandwidth, and then EQ cut with a much narrower bandwidth.

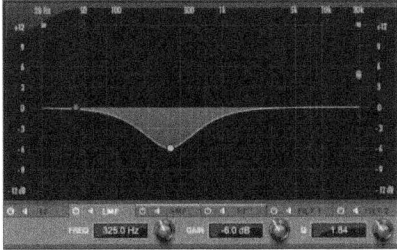 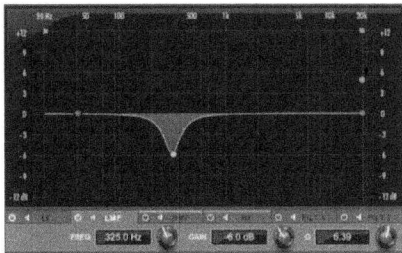

Audio example 90: Low-mid attenuation—flat, wide Q, narrow Q

Here's the difference between a high frequency peaking and shelving EQ. The audio example starts flat for a reference, then switches in a peaking filter to demonstrate a slight boost in the mid-range—it makes the guitar a little edgier. Now, the shelving EQ has the same 3k boost point, but it boosts everything above that, so you hear all the noise and other nasties way up there. In this case we would prefer the peaking EQ, but this will depend on the situation.

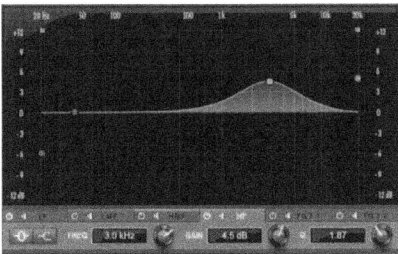 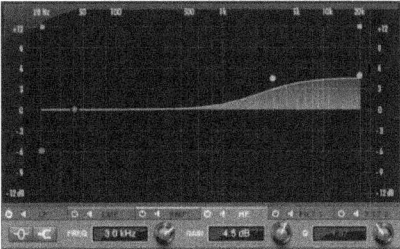

Audio example 91: Hi-freq peaking, hi-freq shelving

How to set an EQ

Some EQ adjustments are pretty straightforward. Need to make a cymbal or tambourine brighter? Turn up the high frequency control. For some consoles the high and low ranges only have boost/cut controls at a fixed frequency, such as 8k or 100Hz.

For the mids, which generally have more controls to work with, many engineers find that it's easiest to over-boost EQ in the range they're targeting, focus on the exact frequency area that needs work, then make the final boost or cut adjustment. If a vocal needs a bit more presence, turn up the high-mid frequency gain control several dB so you can clearly hear the difference. Now turn the frequency select control back and forth to move through the frequencies. Listen for a spot that sounds good for what you need, then pull back the boost to a good level.

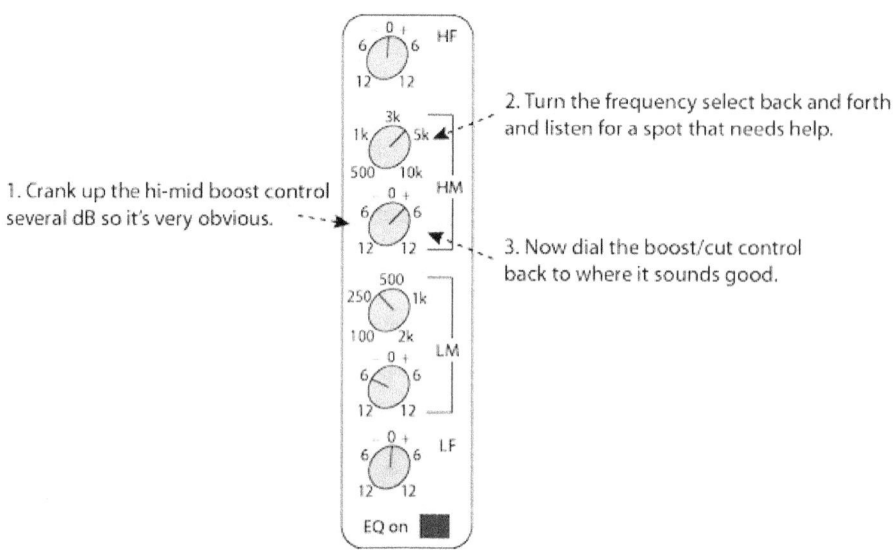

1. Crank up the hi-mid boost control several dB so it's very obvious.

2. Turn the frequency select back and forth and listen for a spot that needs help.

3. Now dial the boost/cut control back to where it sounds good.

For this next audio example, we've looped a guitar part four times. First we're boosting low-mid EQ a lot so we can hear it, then sweep up and down looking for a muddy section. Hear how it gets really boomy and muddy near the end of this first loop? We pull it down below zero to attenuate it 3 or 4dB. The second loop starts bypassed so you hear the original sound, then we cut in the EQ band around the halfway point. Now for the third loop we do the same for the hi-mid band, looking for some natural, nice presence and detail. I like the sound of the picking and strings just under 4kHz, so we bring down the boost so it's up about 2 or 3dB. The last loop starts completely bypassed, then we kick in the entire EQ settings. It might

EQ   155

sound a bit thin and bright by itself, but depending on the song it should fit in nicely and cut through the mix.

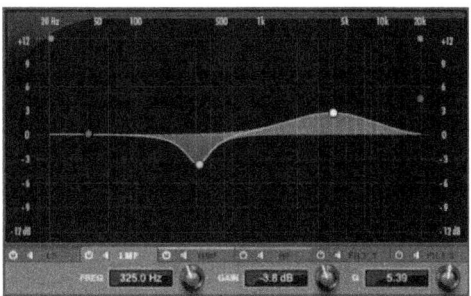

It'll sound weird at first, but with practice you can hear what's "normal" and what's "not what you want".

Here's an entire drum set without, then with EQ on each mic channel.

Listen for the difference between wide and narrow bandwidth attenuation.

Reducing a muddy low-mid range on a vocal to clean it up.

A few pointers to keep in mind

- EQs are used to fix tonal problems such as mic placement coloration and also to change the overall sound of an instrument. If there is a muddy region in the piano mic, and moving the mic doesn't help, then cutting the frequencies at that spot can clean up the sound. If you need something to sound fuller with more bass, then boosting lows can help *to a certain extent.* If there are no low frequencies present, then it does no good to try boosting something that doesn't exist. In other words, you can't make your soprano sound like Barry White no matter how much you twirl that knob.

- EQs should not be expected to compensate for poor quality sound sources or bad miking technique. Don't automatically reach for the EQ—try moving or replacing a mic to see if you can get what you want. I once bought a kick drum mic and it sounded terrible no matter what I did; finally gave up and tried another one. Ask the guitar guy how old his strings are. Send the drummer a drum-tuning workshop YouTube link.

- Most EQ circuits introduce a certain amount of phase shift within the audio signal. This happens due to the way filters affect frequency components in the attempt to boost or cut. After you work with them awhile you'll learn to hear this effect, though for live services it's not really a big deal. Less is better, so try not to EQ too drastically.

- Boosting frequency ranges with EQ adds to the signal's overall dynamic range and level. If boosted too high, it can overload and distort the signal in your console. Keep an eye on your channel and mix meters.

- Most engineers try to attenuate first; humans don't hear a cut as easily as a boost, so this is a more transparent approach to altering a sound. For example, if you reduce a muddy low-mid region, the entire sound will be brighter and clearer, requiring less adjustment in the high-mids.

- Use those low-cut filters. These unsung heroes clean up low-end leakage and rumble, making your mix job much easier.

- If you're doing the same major EQ adjustments on several mic channels, such as pulling highs or lows down 6dB or more, there's a larger issue that needs to be examined in your system. Use channel EQs for minor tweaks, not to correct deficiencies in your room or system setup.

- Once you understand the gist of what an EQ is doing to a signal, you can do a better job experimenting with various sounds and EQ controls.

- If your console runs plugin processing, experiment with different EQ models. They each have a different sound and can really add some unique flavor to each part.

Key frequencies for various instruments

- Kick: body around 80Hz, attack around 2k
- Snare: body 200–300, attack and crispness around 2–4k

EQ 157

- Cymbals: high end shimmer 7k–12k

- Mounted toms: body around 200–300, attack around 5k

- Floor toms: body around 100Hz, attack around 5k

- Vocals: intelligibility 2–4k, presence 4–5k, sibilance 7k or so

- In general, many instruments have a body sound in the 200–300 range, with attack and/or articulation transients in the mid-freq range (2–5k).

- Low-frequency instruments have a bottom end around 100Hz (+/-), though bass guitars go down in the 30s.

- Muddy sound generally happens somewhere between 200–400Hz, give or take depending on the instrument. We often pull this area down to clean up the sound and make it sound less cloudy.

- 500Hz often features a nasal, honky sound that should be removed with a narrow-band attenuation.

Room acoustics *(or, how your sanctuary is ruining your mix every week)*

No matter how much time and effort you put into selecting microphones, getting them placed appropriately, tweaking EQ, and carefully balancing everything, you pull your hair out because it just doesn't sound good in the sanctuary. Some of this may still be due to your learning curve, but the room itself always plays a role. Everybody has heard the difference between a bathroom, living room, gym, and large auditorium, even though they don't know why. The ideal sanctuary distributes sound evenly throughout the space, maintaining a smooth frequency response that doesn't emphasize or lose certain frequency ranges, provides clarity for music and speech, and provides enough ambiance without too much reverberation. Sound like a tall order? Absolutely, which is why a great-sounding church is a rare treat.

I was once hired to finish the installation of a new recording studio. When I walked into the tracking room (where the musicians play during a recording), the door closed behind me like an airlock on the Millennium Falcon. The room was completely dead, no ambiance at all, and it sounded boomy. The problem was too much high-frequency absorption—carpet on the floor, a solid line of sound panels around the walls, and really thick acoustic ceiling tiles. Yikes. Very little to control low frequencies and far too much high frequency absorption. And this was from a professional audio systems design company!

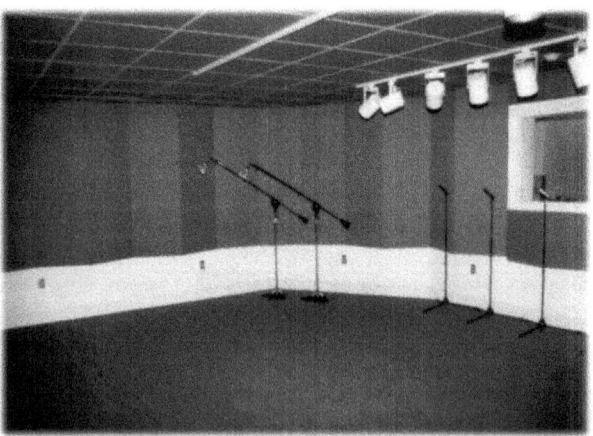

The issues and theories of room acoustics are complex and beyond the scope of this book. I highly recommend that if your church is designing a new sanctuary (or other space) or planning to remodel, they absolutely must hire an acoustician who has done this type of work. What always happens is that the general architect hired for the job doesn't know much about acoustics (though they always claim they do), the church doesn't want to spend extra money, and so they move into this beautiful space and wonder why it sounds so bad. Include acoustics in the initial design before you pour concrete and you won't have to redo and patch the room forever afterward.

Having said all of this, if you're stuck with what you've got (believe me, I felt your pain for years), there are some basic things you can try that will at least make it a little better. Hang with me while we review some fundamental issues and potential solutions that you can do yourself.

Fundamental room acoustics

Once something generates sound in a room, those sound waves propagate (travel) around the room in various ways. The size, shape, and surface treatments of the room determine how this happens, and thus how it sounds.

A propagated sound progresses through three stages:

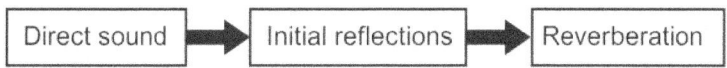

Direct sound is what you hear straight from the source, with no reflections from nearby surfaces affecting it. The closer you are to a sound source, the more direct sound you will hear.

Early reflections are the first waves to be reflected from nearby surfaces such as walls, ceiling, floor, or room furnishings. Unless it's a really big area, they will not be heard as distinct echoes, so you can't sit there and say "Oh, I just experienced an early reflection. That was pretty wild". Your brain, however, is able to detect these various waves and decodes that into a sense of how big the room is. This comes from the delay time between the original direct sound and these early reflections, so this is how you "know" whether you're in an auditorium or small club. A longer initial delay indicates a larger room, because the reflective surface is farther away from the listener and takes a bit longer to arrive. If the delay between reflections is long enough, we hear distinct echoes. This occurs in very large, reflective rooms such as gyms as well as outdoor stadiums.

The major problem with wall reflections is that they destroy your sound. When you hear something that's coming from the stage as well as bouncing off the wall beside you, it combines in a way that affects intelligibility and clarity (remember the phase cancellation discussion earlier in the book?). The more bare wall surfaces you have around the room the worse this problem becomes. My church's sanctuary was built with lots of large bare walls; when someone sang or spoke you'd hear sound bouncing everywhere, creating a muddy mess. Along with making things harder to hear and understand, this posed problems for musicians trying to play together— they heard drums right beside them, then along came the reflections from the other side of the room.

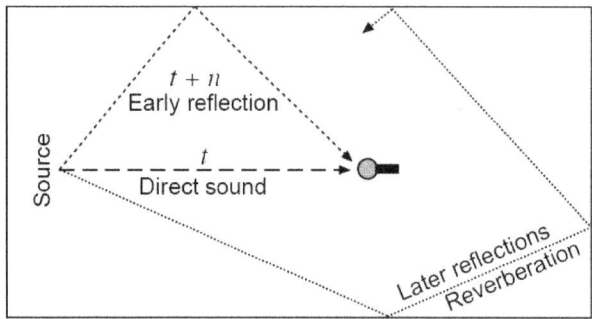

These reflections continue to travel throughout the room and bounce in various directions. If left untamed, they rapidly multiply and develop into a dense sound we call reverberation. If the walls, ceiling, and floor are mostly hard, flat surfaces, you get lots of reflected sound and more reverb (and lots of other nasties). On the other hand, if these surfaces are broken up with various textures and angles, the room will sound clearer and give you a smoother frequency response.

Absorbing sound energy

Imagine you're a stuntman on a movie set and they ask you to jump off a building into the street below. What would you rather dive into, the pavement or an inflatable mat? The asphalt is no good because it doesn't give, so all that energy gets pushed back into you (most likely breaking bones, though I've never tried it  myself). The inflatable mat, on the other hand, is soft and cushiony, so it collapses and absorbs energy from your fall. Sound bouncing around a room is no different. A gym built from concrete block with steel structural beams overhead will be highly reflective since these materials are very rigid and absorb very little sound energy. When you move into a new house or apartment you'll notice that the empty rooms sound echoey and ringy (official technical jargon). Once you add furniture, plants, pictures on the walls, window drapes, bookshelves, carpet, and cats, it sounds much more subdued. There are two main principles here, one of them being that these items *absorb* sound energy. You need soft and fuzzy materials to absorb sound; they're not as dense as brick, concrete, and metal, and so instead of reflecting back into the room, soundwaves get trapped in the material where the energy dissipates. The other principle is that these items *diffuse* sound around the room. We'll get to that in a second.

High and mid frequencies

You've probably seen churches, classrooms, and auditoriums that have sound panels mounted on their walls. Most of the time these are compressed fiberglass panels covered with fabric. This material is an excellent, low-cost sound absorber and has been used for years. I once heard that Noah finally signed off on the purchase order for a huge truckload of these panels when things got out of hand on the ark. The

disadvantage is that they're mostly limited to high and mid frequencies; low frequencies have much longer wavelengths (they're bigger) and require a different approach. Rooms that only have these types of panels on the walls will typically sound somewhat subdued and boomy; the high frequencies that provide ambiance and life to your sound are gone while the low frequencies roam untouched around the room. This requires a balanced approach in terms of materials and installation methods.

Fiberglas panels are available in 1-inch, 2-inch, and 3-inch thicknesses. 1-inch panels are really thin, so they're only effective for high frequencies. Thicker panels easily handle high frequencies, but are also able to absorb a bit farther down in the mid-frequency ranges. Try to use at least 2-inch panels on your walls; 3-inch panels are even better for lower frequencies. You'll have lots of choices of fabric colors to match your room design, and you can even order panels with artwork or photos screened on the fabric (for a hefty price). If you've got folks handy with basic carpentry, you could make these panels yourself. The fiberglass most often used is Owens Corning 703, and you want a fabric that's fireproof rated (Guilford of Maine is a common choice). Build a frame to hold the panel and to wrap the fabric around.

Fiberglas panels should be installed on large bare wall areas, mainly the side and rear walls. Try to cover a good amount of the wall surface for it to make a difference (at least 50%); they can be spaced out from each other a bit or clustered in blocks depending on what kind of look you want. The key is to not leave large blank spots.

Installation is a breeze; you can use construction glue and impaler clips. These metal clips are flat with sharp points sticking out. Attach the flat side of the clip to the wall with screws, then slide the panel onto the impalers.

Low frequencies

Low frequencies are a different animal because they have really long wavelengths and lots more energy than high frequency sounds. To absorb this we mount a semi-rigid membrane (plywood or something similarly stiff) in a frame with airspace behind the membrane. When low frequencies hit the membrane, it gives a bit, absorbing some energy from the wave. Have you ever caught a baseball line-drive with your bare hand? If you hold your hand still to catch it, it stings. A lot. But if you pull your hand back just as the ball gets there, it doesn't hurt so much. You've absorbed much of the energy, like a shock absorber. That's how a bass trap works. The airspace between it and the wall behind acts like a sponge, soaking up more low frequency energy.

The disadvantage of bass traps is that they're much thicker and heavier than standard acoustic panels, so if you want to install them you need to find wall areas that are out of the way. Unlike the acoustic panels, though, which need to be placed according to where sound is reflecting in the room, bass traps can go nearly anywhere

since low frequencies are largely non-directional. You're already familiar with this if you have a sub-woofer in your home theatre system; you can put it anywhere in the room and for the most part you can't tell where it is during the movie.

Another disadvantage of bass traps is that they are expensive, so it becomes impractical to buy enough traps to handle a sanctuary. Most churches don't even bother; the larger the room, the less of an issue this becomes as the problematic frequencies are so low in the spectrum. If your room has drywall, you've already got some low-end absorption built-in as this stuff flexes—just like a trap. If your room sounds boomy, though, consider building traps yourself. Remember, the basic concept is to have a rigid membrane, such as plywood, framed and mounted several inches off the wall. Build your frame on the wall, screw the plywood onto the front, then seal the sides of the frame. Don't add cross-pieces behind the plywood—it must only be attached along the edges so it can flex. For even more absorption attach thick panels of Owens Corning 703 (the compressed fiberglass boards) to the wall inside. The photo on the previous page shows a large trap for a recording studio with framing, insulation inside, and plywood beginning to appear.

The easiest and cheapest way to improve your low frequency response of the room is to mount standard acoustic panels with airspaces between the panel and the wall. You probably won't do this with all your wall panels due to aesthetics, but find some unobtrusive areas and build frames that set 2" or 3" panels off the wall a few inches (more is better). You can also buy offset impaler clips designed just for this purpose.

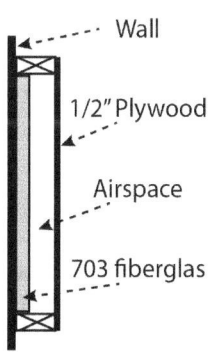
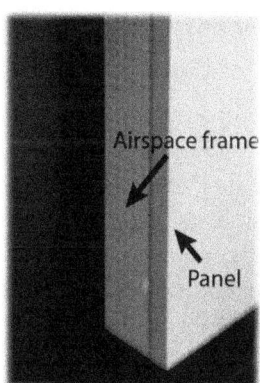

Distributing sound evenly

Absorbers aim to capture sound energy and prevent it from reflecting back into the room. But, if we absorb all our sound we're robbing ourselves of desirable ambiance that makes music come to life. Since we also do not want sound to reflect uncontrollably around the room, we use materials and objects that control these reflections without killing them entirely. This is called *diffusion*, meaning that when a sound hits an irregular surface, the various frequency components in that sound are scattered back into the room in different directions. The trick is to do this evenly

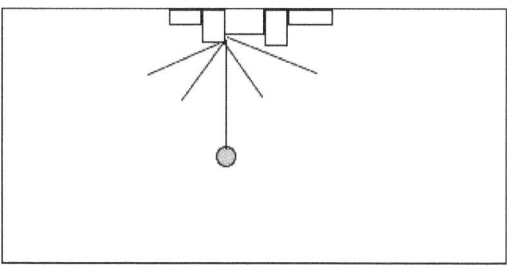

and gradually. Hard reflections merely bounce straight back into the room and cause problems. Diffusion tames these reflections in a way that's acoustically pleasing to the ear.

We diffuse sound by breaking up the flat surface, creating various angles and depths that affect different frequencies. How does this work? Sound frequencies from low to high have different physical "size", known as wavelengths. Low frequencies are really long, as in feet, whereas high frequencies are only fractions of an inch. When these components hit various objects and angles, they'll bounce back in various directions dependent upon how each component relates to the size and angle of the obstacle. Sounds complicated, and much of this is based on some pretty high-level math. But, you don't need to worry about any of that, because we can buy products that take care of this for us. You could also build these yourself once you see some examples.

Diffusers come in a variety of sizes and designs; go check out acoustic manufacturer and dealer websites to get some ideas. They're thicker than standard absorption panels because they must have varying surface dimensions to do their trick. Some models work well on your walls interspersed with regular absorption panels. The wall behind the congregation is an excellent place to install diffusors for a couple of reasons. It's not the most visible surface in the sanctuary, and much of your sound goes right to the back of the room. You could kill it with absorption, but it's often more desirable to include diffusers to provide controlled reflections back into the room, thereby reinforcing your music and overall ambiance.

For my church, I had one of our members build diffusers for us. Not too difficult for someone with the right tools and woodworking skills. We installed these along both side walls as well as at the top of the back wall above the main entry doors.

And then there's this...

Some companies have developed absorption panels that also provide some amount of diffusion. Along with fiberglass for absorption, these products feature an internal panel with punched holes or shaped wells designed to diffuse soundwaves. All of this is covered with fabric. I've used these quite a bit in the recording studios where I work. They're a lot more expensive, but I like the combination of treatments. Even better, these panels can be mounted off the wall several inches, creating an airspace that provides some low frequency absorption. Makes your design job easier as long as you can afford it.

Ceilings and floors

Ceilings come in a variety of shapes, heights, and materials. Flat ceilings are usually either solid drywall or a suspended metal grid. Look up in your office building or classroom and you will probably see a 2x2 metal grid with square, white tiles. Though they call these "acoustic tiles", don't be fooled as they don't do very much. Replace them with ceiling diffusers, which feature a variety of shapes and depths. You don't need these in every grid opening; intersperse diffusers with regular tiles. If your ceiling is solid and fairly low, you've got to do something as sound will reflect directly off this surface the same as it does on walls. Consider flat absorptive panels, diffusers, or a combination of both.

Ceilings with lots of exposed ductwork, support beams, and conduit provide a good amount of diffusion, but hardly any absorption, so your sound will rattle around in the rafters. The easiest solution here is to hang acoustic baffles or blankets. These are thick, cloth covered fiberglass or mineral wool; when sound gets up in the ceiling it gets trapped and soaked up by these baffles. You see this a lot in gyms, but it works for sanctuaries and multipurpose rooms too.

Now, if the ceiling is cathedral-type or very high, it either gets more complicated or pretty simple. The problem with high ceilings is that there may be nothing to reflect sound back down into the room. This sounds contradictory, but if nothing comes back down, you lose a lot of ambiance. Products such as *acoustic clouds* are often hung from ceilings to do a couple of things. They're typically acoustic panels, so they absorb sound. But, you don't hang a solid line of them across the ceiling, you space them out. So the edges and shape of the panels help diffuse sound. Some of them are also designed to reflect sound back down into the room so you get some ambiance. My church has this very high, wood-paneled, vaulted ceiling. Though it's a poor design for acoustics, my solution is really simple—unless they want to spend a lot of money to hang a grid for clouds (thereby hiding all that expensive woodwork above), I won't do a thing to it. Simple, saves a lot of money, and we try to improve the room by concentrating on other aspects.

A quick word about floors. Some great debates heat up when a church is considering laying carpet over a tile floor (or vice versa). The pipe organist and choir director absolutely scream in despair at such a thought. The sound folks plead for carpet. Why all the fuss? A tile floor is very reflective, right? What sounds good with lots of reflections and reverberation? Choirs and, you guessed it, pipe organs. What sounds awful in this environment? Everything else. Footsteps are loud. Kids dropping hymn books on the floor are loud. Spoken word is difficult to understand. Anything amplified through a sound system sounds bad as it bounces around the place. Carpet tames much of this, but it also greatly reduces the reflections necessary for organs and choirs to sound like organs and choirs. You have to decide what your priorities are, slip a few twenties to a couple of folks, and get on with it.

Free treatments

Look around your sanctuary. Are there various decorations, drapes, columns, objects, or whatever that break up the space? Anything that's not flat will provide some amount of absorption and/or diffusion. Cushioned seats will absorb sound, and of course people sitting in those seats. The best part? It's free since it already exists. I like that.

Now let's fix your room

In my early years of college teaching I built a new recording studio in our building, but did not have a budget for acoustic treatment. The room was bare and sounded awful, so one day in acoustics class I brought in a couple rolls of house insulation, all the spare cardboard boxes I could find, and some staple guns. I instructed one team of students to staple the insulation along the top edge of the room while another couple of teams stapled the boxes to the walls in random configurations. The bottoms of the boxes were facing away from the wall. It looked terrible, but it was a rudimentary (desperate?) solution that provided high frequency absorption (the insulation) and low frequency absorption (the boxes). The random placement of various size boxes, stapled to the wall at different angles, provided diffusion to scatter sound waves. Did it sound better? You bet. Did it sound great? Not a chance.

The moral of this story is that at least *something* can improve your room if it follows basic acoustic principles. I now have really nice studio facilities that are far better than those early days. But it took time and intermediate steps along the way. If you don't have a budget for hiring a professional, there are a few interventions that are affordable and easy to install.

For your sanctuary, we need to select materials and products that provide clarity and desired ambiance based on these requirements:

- Mid-to-high frequency absorption
- Low frequency absorption
- Diffusion

Mid-high frequency absorption

Install 2" thick acoustic panels on any bare side and rear walls. Set the bottom edge of these panels roughly flush with the tops of your pews or chairs. You don't need them lower than this because reflections at chair-level will get diffused and absorbed by people and the furniture. But, you don't want reflections bouncing right  into people's ears. So, start just below head-level and extend up from there. Assuming your panels are 4' high or so, you might need another row along the top depending on how tall the walls are. Panels can be flush against each other or spaced out several inches, depending on the look you want and how much money you have (it adds up quickly). Don't cover the entire wall; this will soak up too much sound, giving you a dead-sounding space that nobody will enjoy. Moderation is always good advice.

Another idea is to hang thick drapes, flags, banners, or old sweatshirts on your walls. Not as good as a real panel, but it's a step in the right direction and you'll definitely hear a difference.

Low frequency absorption

Look for some wall space where you can build a bass-trap using the plywood technique I described earlier. These should be several inches thick, so this will limit your options. You can build individual 4'x8' traps, or perhaps line an entire wall with 4'x8' sheets oriented horizontally end to end. If you set these at floor level you've got yourself a free shelf—people can stack bulletins, Bibles, or even stand on it to hang Christmas wreaths over the windows. Compressed fiberglass boards are cheap, so put at least one 2" or 3" board inside these frames against the wall to improve their effectiveness. Even though you can purchase individual bass traps (they come in various shapes and sizes), this will usually be too expensive for such a large space. If you can't do any of this, don't worry. Most churches don't do it either. Mount some of your acoustic panels off-the-wall a few inches, creating an airspace behind them, or you can install thicker panels.

Diffusion

It's really critical to not over-absorb sound in the room. Many churches plaster their walls with fiberglass panels, but this kills the room. You really need to include diffusers to preserve a nice ambiance for music.

The diffusers shown here are located along both side walls in my church. We've also installed them over the tops of the entrance doors and windows on the back wall behind the congregation, combined with absorption panels. This helps control ambient reflections back into the room. A good rule of thumb is to put diffusers along the side walls near the front at people-height (sitting and standing), then transition to absorption panels further toward the back of the room. This reinforces music and singing where the people are, but reduces corner and rear wall reflections back into the room.

Another idea is to alternate patterns of absorption and diffusion, including a panel of diffusion every so often so you don't soak up too much of your ambiance.

You'll also get some diffusion from various objects, furniture, and architectural elements in the room (such as those ornate columns and carvings in old cathedrals). The goal is to avoid large bare, flat walls and ceilings at all costs.

This is an early design idea for the left side wall of my church's sanctuary. You can see diffusers (striped sections) from the middle toward the front (right side of the diagram faces the stage), then the back section transitions to absorption panels—thick 3" panels for some low frequency control. The front two sections are where the band is located, so I've put the diffusers to help smoothly reinforce acoustic sound into the room with absorption down low to control boominess which typically builds up here. It didn't exactly end up this way, but you get the idea.

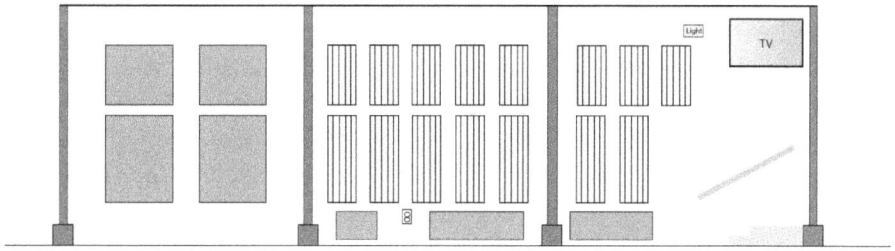

What to do with all of this

Though I've tried to explain various options to help you with your church, there's no way I can be specific. The basics are fairly straightforward: balance absorption and diffusion, making sure you treat the entire range of frequencies. Tame surfaces that reflect sound back into the seating area and don't leave large patches of bare walls. How much of each depends on the room size, shape, configuration, surface materials, type of use, and budget. There are lots of books and online resources to guide your way. For example, acoustic product manufacturers provide lots of theory and application suggestions on their websites, including photos of facilities showing acoustic treatment. Your team can probably do quite a bit on your own with some research, but at some point it's best to hire an acoustician or installer who really knows what they're doing, saving you money and frustration in the long run.

Audio fundamentals

The heart of everything we've discussed involves fundamental acoustics and electrical signals. Understanding these elements helps you understand how we capture sounds, route them through equipment, process them, and listen through speakers. You'll run into situations where a solid understanding of acoustics and audio signals can help you determine possible causes and potential solutions. We've covered some of this here and there already, and this section only attempts to provide a foundation; there are many excellent books available that describe these concepts in greater detail.

Sound waves

How do sounds actually get from one point to another? If you see someone moving their mouth as they look at you, how does that covey anything meaningful such as speech? When we hear someone playing a musical instrument, how can we tell that it's a piano and not a snare drum? The answer lies in the vibration of air molecules; when an object vibrates, such as vocal cords, piano strings, or a chainsaw, it changes the atmospheric pressure around it. These are relatively minor pressure changes— nothing like what happens when a hurricane blows through. But they are enough that when the vibration follows a particular pattern, it can convey information when received by our ears or by a microphone.

What's a pattern? Depends on the vibrating object. If it's a drumstick hitting a drum, then the sound pressure variations are fairly random and sound like noise to us. If it's a flute playing a note, then the sound waves follow a repetitive cycle—the particular waveform pattern of a flute repeats over and over until the playing stops. This is called a periodic wave, and all musical sounds feature this characteristic.

The sound itself is also distinguished by what the waveform looks like if you graph it out. Sound generated from a flute will set up a particular pattern of fluctuations different than that of a snare drum. These waveforms are built from simple sine waves—lots of them. You remember sine waves from geometry class, though you probably had no idea what they were used for. Now you have a practical application; sine waves, the most pure, simple form of vibration, combine in varying patterns to create complex sounds—instruments, noises, and speech. When these patterns repeat over and over you get a musical sound; when they are random you get noise.

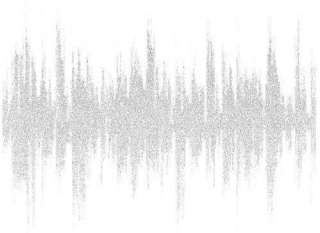

A complex musical waveform

In audio equipment such as consoles and processors there is no physical object vibrating. When an acoustic sound is transduced into an electrical signal inside a microphone, the pressure variations then correspond to resulting changes in voltage. An increase in pressure equates to an increase in positive voltage, and a decrease in pressure is represented by negative voltage. Thus the electrical signal version is analogous to the original acoustic sound (thus the term *analog* in the world of audio). This can be seen by graphing a simple sine wave—the positive curve going up represents the original acoustic pressure increase, and the curve going down into negative territory comes from an acoustic pressure decrease. This constantly fluctuates around the zero point, which is equilibrium (no signal), just as the AC voltage in your house constantly fluctuates between positive and negative. With audio signals, however, the fluctuation is much more complex and interesting compared to the plain 60Hz repetitive cycle in your power lines.

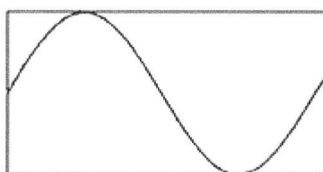

Simple sine wave: upward curve is positive voltage, downward curve is negative

As you probably know, these vibrating patterns must oscillate (vibrate) between 20 and 20,000 times per second to be heard by humans. This is known as *frequency*, number of cycles per second, which roughly equates to our sense of pitch (high or low). Translating this to musical terms, particular pitches are based on specific frequencies, such as the standard tuning note of A = 440Hz. A musical octave can be heard every time the frequency of a sound doubles, meaning that if you play a 40Hz tone, then double that to 80Hz, it will sound like the same pitch, only an octave higher. Audio equalizers are built with this in mind, where graphic EQs feature filters set at octave intervals, third-of-an-octave intervals, etc.

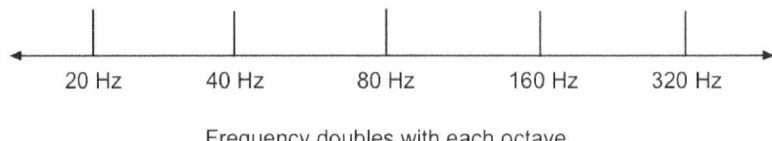

Frequency doubles with each octave

Different frequencies have different wavelengths; low frequencies will have long waveforms (30 feet and longer) while higher frequencies possess much shorter waveforms. As we discovered in the previous section on sanctuary acoustics, this is important when you're trying to improve how your room sounds.

How much the variation moves from equilibrium (amount of pressure change) is referred to as *amplitude*. This is what gives us a sense of loudness; pluck a guitar string harder and it sounds louder. In electrical signals running through your equipment it means how high or low your signal strength is; moving a console fader up increases amplitude, moving it down decreases it. If signal level is too high for a circuit it will distort; all audio equipment (including our ears) has a limited dynamic range, or the capability of handling amplitude ranges. If a signal is too low, we get an increase in the amount of noise that is heard in the sound, so the sound engineer must balance signal levels pretty closely to get the best, cleanest sounding signal possible.

A few more tidbits about audio signals

Wonder why old recordings seem to lack that luster and clarity of newer recordings? Older recording equipment lacked the frequency response of modern technology, meaning higher and lower frequencies on either end of the audio spectrum simply could not be captured and reproduced. Most obvious is the standard "telephone" sound, which we now mimic as a vocal effect by rolling off the highs and lows, leaving only the mid-range. Knowing the frequency ranges of various instruments and voices helps you to better find EQ settings that fit that particular part. We select different microphones for certain miking situations based on their individual frequency response and sound—they're intentionally made to sound unique. So, you'd most likely put a large-diaphragm mic on a kick drum to capture its low frequencies, but not on a flute which has very little low frequency energy.

We mentioned that signal amplitude refers to how much juice is flowing through your circuits. Over-drive your microphone preamp and it'll distort. Crank up too much EQ and you can overload the EQ circuit. Main amplifiers can be overdriven into distortion. If you set signal levels too low, however, you get a lot more noise added to your sound because the audio signal isn't loud enough to cover it up. Connect a piece of gear into another that's not matched correctly and you'll have problems getting the levels to work. Ever notice if a piece of equipment has a level switch on the back, say between +4 and -10 or -20? That's there for a good reason, and you need to know which level your overall system is running so you can purchase and connect equipment that will work correctly.

We've discussed phase issues that exist acoustically when you use multiple mics on stage, and the results show up when mixing those signals on the console. The most common situation is miking a drum set. When you're using several mics at once on a drum set, start listening to one mic at a time, gradually building the drum mix (it's easier to hear with good headphones). At some point you might notice that the kick, snare, or something thins out. You just added a mic that's in adverse phase relationship with the mic on that particular instrument. Either move one of the mics a little (might not take much), or try flipping the phase button on one of the mic channels on the console.

We can run into phase problems electronically as well. Take a recorded track in your software, make a copy of it to a new track while adding a short delay, then mix them together and listen for how the sound thins somewhat (depending on the delay time). Incorrectly wired cables will certainly cause problems. For example, mic cables have three wires inside, two for signal and one for ground. The two signal wires must be connected to the correct pins, and if these are reversed at one end of the cable you've got a complete 180 degree polarity shift. That's a problem, so you should check your cables with a cable tester to be sure.

Human hearing

We won't get into a complicated discussion about how the brain deciphers sounds transduced by your ears. For now I'll focus on a few key aspects that might help explain various things you run into along the way.

The main concept is that human hearing is non-linear. This means that you don't hear what you think you are hearing. In other words, there are various issues involved in how we perceive sounds. For instance, if you double the wattage of a power amplifier, you would assume that the volume would double. Not so. The actual perceived increase is only 3dB—barely noticeable. It takes 10 times as much amplification to make it sound twice as loud. Remember studying logarithms in school? This is where it comes into play, where changes in loudness (and pitch) is logarithmic and not linear. Later I'll briefly introduce you to decibels and using logarithms to calculate various sound and signal levels—don't freak out, it'll be okay.

There are also differences in sound depending on how loud we listen. If you listen at lower volumes, then you hear less bass and treble frequencies. If you crank it up very loud, you will hear much more bass and treble than usual. This is particularly critical when recording and mixing in a studio. If you engineer your recordings based on these listening levels, then they will sound quite different when played back in various situations. There is a graph that shows us how this works (Fletcher-Munson), and the suggested solution is to listen at about 85dB SPL. There are phone apps available that measure sound level, which would also be good for checking how loud your overall mix is on a Sunday morning.

One other issue to remember is that our ears can be "overloaded" with excessive volume. Not only will this cause permanent hearing loss over time, but there are immediate implications as well. Your ears will quickly tire and lose perspective the longer you work. If you quit for the day and come back later, you'll notice how different your recordings and stage mixes sound. You should take breaks regularly during long sessions. If you listen to very loud levels for any extended period of time, you will also begin to hear distortion at some point, but this is actually happening in your own hearing system. The concept is exactly the same as what happens when you overload a power amplifier (or any audio component). The signal hits maximum dynamic range and any level over that point gets clipped. This essentially creates a square wave, adding a host of unrelated sine components. The resulting waveform is different and adds noise that is usually unwanted and unpleasant. Our ears do the same thing. Early in my career I was working with a record producer who liked to monitor at really high volumes. It wasn't long before I was hearing distortion in the monitors, but it wasn't from any signal levels I had running through the console. It took a moment to realize it was all in my head, so to speak, but in a very real way. Knowing I had set everything up okay before this kicked in allowed me to complete the session without worrying about what was being recorded, but it was a weird experience.

Perception of direction

We talked briefly in the stereo microphone technique section about how humans hear direction of sound. Using differences in amplitude and time of arrival between left and right, we can tell where sound sources are located. This is also used to create a sense of direction when mixing recordings. Pan potentiometers and digital delays follow these two principles when placing images in various stereophonic locations. Different stereo miking techniques follow these principles in capturing a sense of stereo space from the stage source, so if you're mixing live in stereo or planning a recording, keep this in mind.

Perception of space

People get a sense of the room they are listening in through such cues as direct sounds and reflections. Once a sound is generated, the listener will hear the direct sound straight from the source, then begin to detect early and late reflections from nearby surfaces. These are not distinct echoes, but very closely-spaced reflections in the millisecond range. The longer it takes for the brain to detect an initial reflection after hearing the original direct sound, the larger the room is perceived to be. Think about it—if you're in a large room the walls are pretty far away, which means it takes longer for those first reflections to arrive where you're sitting. So what? If you're making a recording, the reverb settings you use can recreate a large concert hall or a small jazz club. We'll also use this principle when trying to improve room acoustics by taming those reflections before they get out of hand.

Decibels & logarithms

In the world of audio we need to measure acoustic sound and signals running through circuits. We need to know how loud a concert is, how many speaker cabinets and amps will be needed to run the show, where to set our meters on the console, and find out the before and after affects of an increase or decrease in sound or signal level.

So, how do we go about measuring signal levels? When you look at a console you'll see meters labeled in units called dB. So when the vocal peaks at around 2dB on an analog console meter, what exactly does that mean? Nothing by itself. Decibels are meaningless unless they are referenced to something that has value. Unlike a yard, meter, or mile, which have very specific quantifiable values, decibels have no value in and of themselves. What we do is compare them to values that are based on specific measurements. Huh?

In acoustics, we use a dB SPL meter to measure how loud a concert is. The result, say around 110dB SPL, indicates that the concert acoustic level is 110 decibels louder than the threshold of hearing (0dB), which is the lowest point humans can detect sounds. In electronics, we measure a vocal signal running through a channel by comparing the current voltage (or wattage) to an established standard reference level for audio equipment. So, relative to what has been established as 0dB for voltage readings (remember unity gain?), the vocal signal is occasionally higher by

2dB. Decibels are essentially expressing a ratio between two known quantities, providing useful information that can be employed in a variety of situations.

Why use dB? Decibels are logarithmic, which is useful to us in a couple of ways. First, the range of acoustic intensity is tremendous—lots of really big numbers that engineers don't want to use in daily life. The same is true for voltage readings and such. It's much easier to find a unit system that reduces this to a more manageable way of working with changes in SPL or signal levels. So, even though the difference in sound pressure between the threshold of hearing and a space shuttle launching in your driveway is almost beyond comprehension with lots of digits and decimal places, we can easily say that the difference is around 150dB (give or take). Just remember that every 10dB increase is about equal to a doubling of perceived volume, so small numeric increases in dB actually represent significant changes in intensity or level as you go up the scale (that's the log part of it). Every time you double your sound system amplifier wattage you're not getting a sound that's twice as loud, but rather a small increase in perceived volume. That's why it takes racks and stacks of amps to run a large sound reinforcement system.

The other reason to use logs is that we hear logarithmically. Remember the fact that every time frequency is doubled it sounds like an octave higher, no matter where you are on the spectrum? The actual quantity of frequencies between an octave of 20Hz–40Hz and an octave of 10,000Hz and 20,000Hz is very different. The bandwidth is 20Hz in the lower octave and 10,000Hz in the higher octave. However, each octave *sounds* like the same interval, whether it's in the high or low range.

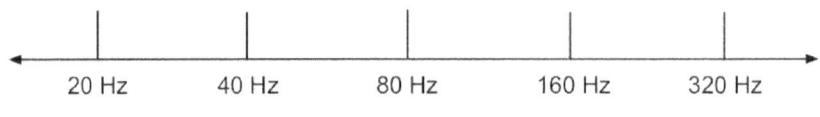

20 Hz   40 Hz   80 Hz   160 Hz   320 Hz

Frequency doubles with each octave

Logs provide a system of measurement that allows the use of reasonable mathematic figures and comparisons that match our non-linear perception of hearing. If you recall from math class in high school (yeah, right), a log of a value (X) is the number which, when applied to 10 as an exponent, produces that value (X). For example, the log of 100 = 2, because two 10s, 10x10, equals 100. These days we just use the formula and log functions in our calculators. Here is one such formula for calculating a dB increase in amplifier output, where we are comparing the original power level (P^{ref}) to the new power level ($P^{measured}$).

$$dB = 10\log P^{meas} / P^{ref}$$

In case you missed it earlier, the decibel system comes from the original telephone pioneer Alexander Graham Bell. Telephone systems actually used the unit *bel*, but it's too large for audio, so we measure in increments of one-tenth of a bel, or *decibel*.

Different types of dB measurements

Decibels are difficult enough for novices (and many professionals) to comprehend, but it gets even more complex due to the different decibel applications and references. Essentially decibels are used for the following applications:

- Measuring sound pressure level (or intensity) to see how loud it is (acoustic measurement).

- Measuring signal level (electrical measurement).

- Ensuring optimum signal levels during live reinforcement and recording sessions.

- Calibrating equipment.

- Measuring changes in level (acoustic or electronic).

You should understand that when we use a dB indicator, it is based on some other physical measurement such as voltage, wattage, or acoustic pressure. The equipment measures the physical quantity, then we convert that into dB indicators for ease of use. Decibels always provide a comparison between two physical quantities, whether it's an established reference level or the difference between two current measurements. Here is a brief outline of the more common applications.

Measurement applications

dB SPL

Sound Pressure Level (acoustic). We measure pressure change of acoustic waves in comparison to the threshold of hearing (zero). We can also measure sound intensity, which uses a different physical measurement. Higher SPL levels correspond to higher perceived volume.

n dB SPL above the threshold of hearing

$$dB\ SPL = 20\log SPL^{meas} / SPL^{ref}$$

dBu (dBv)

Pro audio equipment voltage. This is the most common measurement for audio signals, and the 0dB reference is equal to 0.775 volts. In your equipment, however, 0dB on the meter actually equates to +4dBu, or 1.23 volts, if you measure with a voltmeter. Professional equipment runs higher "behind the scenes" for better audio quality, thus the reason why 0dB VU is actually +4dBu. Now I'm sure that makes a lot of sense.

n dBu above reference of 0.775 volts

$$dBu = 20\log V^{meas} / V^{ref}$$

dBm

Pro audio equipment power. You'll see this specification in many equipment manuals, but it's not used much from day to day. Originally based on telephone system equipment that ran on 600ohm loads, dBm is used for power measurements (wattage) and not voltage. It's only good when measuring in 600ohm loads, however, and you don't use this for calculating power amplifier output levels.

n dBm above reference of 0.001 watts

$$dBm = 10\log P^{meas} / P^{ref} \text{ (P = power, or wattage)}$$

dBFS

Typically referred to as 0dBFS (full scale), this represents maximum level in a digital system. If you look at meters on analog recorders and mixers you'll see 0dB just past the halfway point with readings beyond this, in the red, up to +3dB. 0dB is optimum, but not maximum. On a digital system, however, 0dB is as high as it goes before it runs out of bits to encode the signal, resulting in immediate, harsh distortion that will ruin your day.

Useful tips to keep in mind

- Doubling of power (wattage) only results in a 3dB increase—not much in perceived volume. So if you just convinced your spouse to replace your 50W guitar amp with a cool 100W amp, you will be quite bummed.

- Doubling of voltage results in a 6dB increase. This comes into play with console levels such as faders and mic preamps.

- Doubling of distance results in a 6dB decrease. This principle comes in mighty handy on stage when miking multiple instruments. Move the sources farther apart along with their mics; if a mic is moved twice as far away from another source, the leakage into that mic will drop by just less than 6dB or so. You'll get a cleaner sound overall. Conversely, if you place a mic closer to its source, you won't have to turn up the fader as much. This will reduce your possibility of feedback, which is always a good thing.

Here's an example of how the math works (you'll have to experiment with your calculator to make sure you know how to use the log functions correctly).

If a shuttle launch produces a sound intensity of 20 watts/m², use the standard formula and plug in the reference value of 10^{-12} watts/m² as follows:

$P = 20$ watts/m² (the shuttle)

$PR = 10^{-12}$ watts/m² (threshold of hearing)

$dB = 10 \log P/PR$

$dB = 10 \log (201/10^{-12})$

$dB = 10 \log 2(1+12)$

$dB = 10 \log 213$

$dB = 133$

Another example:

Measured sound pressure level = .075 dynes/cm²

Reference level = .0002 dynes/cm²

$dB\ SPL = 20 \log .075/.0002$

$dB\ SPL = 51.5$

Here's how to tell the dB increase with your new guitar amplifier:

Original guitar amp was 50 watts

New guitar amp you got for your birthday is 100 watts

$dB = 10 \log 100/50$

dB increase = 3.01 (pretty sad...)

Note that with wattage (power) measurements the formula uses 10 log, whereas with voltage and sound pressure measurements 20 log is used. This is true for both acoustic and electrical calculations.

Tools to measure SPL and signal level

Meters on the console or recorder

We've mentioned audio meters from time to time. There are two primary types, and they provide a visual indication of signal level. However, they operate quite differently and will affect your decisions accordingly.

VU (Volume Unit). These have the little needle that swings over toward the right. VUs provide an averaged value that corresponds to our perception of volume. It will not show all the peaks and dips, the same as the fact that our ears don't register very brief bursts of signal (transients). You can overload something without realizing it, so consoles typically also provide *peak* LED indicators that will flash when you're pushing close to the danger zone.

Peak Reading. These are usually step graph indicators or simple LEDs. They are designed to respond to any instantaneous signal change, and will therefore show all peaks and dips in levels. Sometimes they will also feature a "peak hold" function, meaning it will keep the highest indicator lit for a little longer so you don't miss any transients along the way. Note, though, that some LED step indicators can also be set to VU rather than peak, so know the difference.

VU　　　　　　　　　　　　Peak

Sound pressure level meter

SPL meters measure acoustic sound pressure levels as we described earlier. They actually measure the changes in pressure using $dynes/cm^2$, then convert this to a decibel reading that makes sense to engineers. They're pretty easy to use, but you have to understand the weighting network system to use them properly.

If you read up on Fletcher-Munson curves, you'll know that this graph shows that we hear low, mid, and high frequencies differently at different volumes. To get an accurate reading on an SPL meter, you must adjust the internal weighting network for the general volume level you will be running in your situation. These are referred to as A, B, and C weights with the following applicable SPL levels:

A: Referenced to 40 Phons (levels of 40dB or lower)

B: Referenced to 70 Phons (levels in the 70dB range)

C: Referenced to 100 Phons (levels of 100dB or higher)

To find out how loud your service is running Sunday morning, hold up an SPL meter. Set the weighting to match the levels you're getting. 100dB and above is quite loud...just so you know...at some point those folks in the congregation who constantly whine about the volume will begin pulling out their own iPhones and wave them in the air capturing SPL readings. That's when the game is up.

Other audio meters

There are several other devices used to measure audio for various applications. Standard multimeters can be used to directly measure voltage, current, or resistance in an audio circuit. Specialized audio tools can measure dB comparisons, frequency response, noise levels, impedance, etc. These can range from $10 for a cheap multimeter to several thousand dollars for high quality audio maintenance equipment. Many companies have developed really good audio tools that run on your phone, and they're very affordable. They use either the built-in mic or an external mic to capture sound and provide the same processing and analysis you find with hardware tools. You can then save the results and email them to your system designer. Amazing stuff.

THE CONSULTANT'S REPORT

So, how many issues did you catch from our opening story? Was there anything you weren't sure how to fix? Let's go through the list in the order we ran into them.

First, it's common to not have anybody who knows audio. Welcome to the really big club. Almost everyone who runs sound for their church does it out of interest and dedication, though you can certainly learn enough to be quite dangerous. I also really endorse the idea of including young folks who want to learn this stuff. After all, they're the ones who will run your church someday. Might as well help them learn how to do it right.

Now, the video screens you noticed going through the front lobby. This is a really nice feature, but it requires a quality audio feed where people can hear what's going on in the room. Remember the discussion on using aux sends to create independent mixes? Use this to feed the video system. And go out in the hallway and listen to it. Regularly.

All independent mixes you send to various destinations will sound dry since there is no natural reverberation on them. The set up can be a bit tricky depending on your gear, but you can run *another* aux send on your console channels to a separate reverb processor. Turn up this aux send on all channels you're sending to the video stream. Patch the output of this reverb so that you can get it to the specific aux send mix that's feeding your video screens; one way is to send the reverb output into a spare channel (but don't send this channel to the main sanctuary mix). Turn up the aux send on this channel that feeds the video distribution and blend it with the other parts. So, you could run aux 3 for your video mix and use aux 4 to add reverb to this mix. Aux 4 on all channels goes to the reverb, which comes back into a spare channel where you turn up aux 3 to send this verb to the main video mix. This sounds complicated, but read through it a few times and just break it down step by step.

Evaluation of the sanctuary

A wider room that's not so deep is an effective design so you don't have people stuck way in the back. Many newer sanctuary designs are fan-shaped, with a seating area arranged in an arc. Rectangular rooms, however, feature directly opposing walls. Sound waves will bounce back and forth between these unless something stops or redirects them, meaning sound absorption panels and diffusers.

Carpet is okay, but it tends to soak up high frequencies (and nothing else), meaning you lose some life and ambiance in the room. A mixture of carpet and tile/concrete is a good solution, though a hard surface is louder when people walk on it and drop things (you know, coins, pens, babies).

The walls here are bare, meaning plain drywall that directly reflects sound energy. Soundwaves bounce from wall to wall, resulting in destructive early reflections and a boomy room that lacks clarity. Diffusers and absorbers are essential.

Plush seating absorbs sound, which is a good thing for amplified music. For more traditional churches with choirs and pipe organs, though, hardwood pews and other typical architectural components provide a more natural, live, blended sound.

Much like the bare walls, a flat ceiling is terrible. Those cheap acoustic tiles don't do much. Since there's going to be space with grids and beams above them, open that up. A combination of vertically-hanging baffles and mostly-horizontal cloud panels will help reflect some sound back down in a controlled way while trapping some sound up above. This will reinforce congregational singing while eliminating direct reflections (and that nasty phasing problem we talked about over and over). You could also install diffuser panels that fit these types of grid systems, interspersing them with regular tiles. My church has the opposite problem with a very high cathedral ceiling. There's really nothing to reflect sound back into the congregation, so everything gets lost. It's too bad because people don't really hear others singing; it sounds anemic.

Nothing else stands out about the room, assuming the speakers are properly installed. And yes, that would be a really cool pulpit in my church.

Set up

The loud BOOOMM when powering up the console happened because the backstage amps were already on (most likely by Madison in this case). When you fire up the console it sends a pulse out to the amps. This can damage stuff, so remember the power up sequence: amps on last, amps off first.

Madison should not be asking where to plug the guitar in—all instrument and vocal mic channels should be standardized, labeled, and charted backstage and at the console. Otherwise you're just inviting confusion and lost time trying to trace why a mic doesn't work.

When you noticed only a couple of aux sends being used on the console, this is an indication that they don't know how to set up independent mixes for things such as the recording and building video feed. And yes, personal mixing systems are fabulous and a zillion times better than using monitor wedges on stage. Try to switch to something like this at all costs. Also note that you had to ask Jonathan what each aux was being used for—always label your console.

All mic channels were routed directly to the mix bus. This works, but if you route all vocals to a sub-group and all drums to another sub-group you have more control. You can adjust the overall vocal level in the mix with a single fader. You can put a compressor on the group. You can mute the entire rowdy bunch if they get on your nerves. Control is a very good thing.

By the way, this church had already converted to a digital console, but analog boards are just fine no matter what you read in magazines. Digital models offer some really nice features and many churches are making the switch, but there's nothing inherently obsolete about an analog board.

The CRAAACCKK when plugging microphones in is because the console channels are on. This is bad because it can damage things. Always mute the channels when connecting, moving, or disconnecting mics. Note that this does not stop signal going through any pre-fader aux sends and direct outs that may feed your monitor system, so you'll still hear the pops and cracks there. Turn down the mic preamp to take care of this. Train your guitar friends to ask before plugging and unplugging their direct boxes.

Your heart should sink when you see a guitar dude hauling in any type of amp. Notice in our story that the electric guitar was constantly too loud and uncontrollable. This is because the amp was big, loud, and facing directly toward the congregation. Try to get it turned down, spun around toward the side, put back behind a barrier somewhere, or replaced with an amp simulator.

Don't hang a microphone straight down over an amp. It's now pointing toward the floor, which is not what you're trying to capture. The side of the microphone is facing the speaker cone, which gives you a very colored, unnatural sound. Put it on a short stand in front or buy a guitar amp mic designed to face sideways.

Their drum miking technique needs some help. Put the kick mic inside the shell, never out in front of the head. Snare and tom mics should be a few inches above the head, slightly inside the outer rim. Too close and they sound boomy, too far away and you get more leakage from other drums. Drum shields are really nice and provide everlasting salvation for keeping drums out of all the other mics. But, they're reflective and will bounce sounds back into the wrong ends of the mics. I like to put absorption panels inside to control this a bit. The 4' high models are not tall enough and can allow cymbals to ring out over the tops. Not good.

They never got signal from the acoustic guitar because he plugged out of his guitar into the output of the DI. Outputs always go to inputs...always.

Sound check and walking around

The sound booth is generally well-organized, though cables need to be bundled and labeled. Otherwise it's too easy to catch one with your foot, and you can never remember where each cable goes. Equipment manuals, batteries, and so on should always stay in the same spot and be handy when needed. A raised platform is extremely nice because you can't hear anything if you're stuck behind a bunch of people standing, and you don't really want to stand the whole time either. Always install wireless antennas up as high as you can, so their installer seemed to know what they were doing.

The room sounds boomy because of the bare walls. You'll get much greater clarity, intelligibility, and just plain better sound if you acoustically treat the room.

Don't let your vocalists pick up a random mic each week. You need to set gain, EQ, and maybe compression for each voice because they're all different. Clearly label the mics up front and get them into the habit of using the same one each week. Digital consoles are really helpful because you can save presets for each person. Just keep in mind that even this will need some tweaking from time to time; each service is somewhat different. I remember a theme park show manager who had someone set the console at the beginning of the season, then literally taped everything down so it couldn't be changed—ever.

Hum in the system comes from a variety of things. In this case it's probably a direct box on stage that needs its ground lift switched. Just flip it to whichever position lessens the hum.

I'm not sure why Jonathan struggled to get faders close enough for a mix after the group started playing. Seems basic, but if you run sound every week you should have a pretty good idea of where the faders and mic preamps are usually set. Also realize that mic levels can vary drastically depending on several factors: Does the band know the song? Is it early in the morning and they're just warming up? Are they working something out amongst themselves and stopped singing momentarily? Pay attention to what they're doing before you start making changes, meaning look up and see what's going on. Too many sound guys stare down at the board all the time. I have no idea what's so fascinating about that.

Wonder why the horns are ripping your face off? They're pointed directly into a condenser mic that's located straight in front of them, and they're also firing directly toward the vocalists (and their mics). Turn them away from everyone else and raise their mic higher, pointed down a bit toward them. Get it out of the line of fire, especially for trumpets. Or get rid of the mic entirely if you can hear the horns acoustically in the room.

As the worship team continued running through their songs, Jonathan and Madison tuned them out as they worked on other preparations for the service. This is understandable, but not good. You need your sound person to focus on, get this, sound. All the time. He lost valuable time that could be used to tweak the sanctuary mix, mixes around the building, recording levels, and look for potential problems.

During the service

Be ready to start the service. You're now officially on duty and must concentrate totally on your job. Get mics turned on, levels ready. By the way, if you miss a mic cue, don't just turn it on mid-sentence or song phrase. Pull the fader down, turn on the channel, and quickly but gradually turn the fader back up (remember where it was). This is much less jarring and provides a smoother transition. Of course, even better is to not miss a mic cue because this is one of the most disruptive things you can do in a service—far worse than having an iffy mix.

One more mic on/off tip: Turn mics on *after* they've picked them up from the stands, and turn them off *before* they set them back down. This eliminates thudding in the system as the mic bumps against the clip.

Feedback. Again, pay attention constantly. In this case the ringing was a low frequency tone, so turning down the high frequency EQ did nothing but destroy the sound of the vocals. Learn basic frequency ranges and what they sound like. Maybe this is a system-wide issue that can be resolved with an EQ on the mix bus, or maybe it's just a certain mic (and certain vocalists whose particular voices will trigger various frequencies in the room).

The electric guitar again. Boy, I hate amps in church. They're certainly from Satan's toybox.

Your building feeds must be checked constantly. Listen to make sure that 1) the speakers actually work and 2) the mix sounds decent. Use separate aux sends and remember that your sanctuary mix will sound nothing like it; at least get it close listening on headphones at the console, then walk around from time to time to hear exactly what it's like out there. And yes, we did finally replace all our cheap building speakers at my church, in case you were still wondering.

Lapel mics are finicky about placement issues. Too often they get rubbed against a coat lapel or necktie, and when the person turns their head the mic stays put, thereby losing significant volume and changing the tone of the sound. If they look down to read it gets a lot louder, probably boomier, with a good chance of p-pops. Headworn mics, those cool-looking, nearly invisible mics that sit off to the side, solve most of these problems because they follow the head. All of these, though, will suffer when the person is reading or standing at a podium. The hard surface in front reflects sound back into the mic, causing them to sound like a space alien. Put something soft on the pulpit to absorb some of this, and show them how to angle the Bible or paper away from the mic as they hold it.

Again, the Pastor in this case sounds boomy in the room due to the poor acoustic treatments (or lack thereof). His mic is cutting out due to wireless issues. Might be a low battery, but sounds more like a reception issue, meaning you've got a dead spot (or spots). Raise your antennas pretty high, move them around a bit for different angles, and make sure the transmitter (the little box that clips onto a belt loop) is not stuffed deep inside his money roll pocket. All wireless mic manufacturers have good advice on their websites for dealing with these types of issues. Good luck and God Bless.

Do all of your recording via aux sends, not the main mix bus. Create a custom mix and consider some compression to help control level fluctuations a bit. Always *normalize* the recording before you post it or make CDs. If your church sells, or even gives away for free, recordings that include music, make sure you're doing it legally.

Post-service music from the iPod. A great idea, but smooth transitions are the goal here. Don't just turn it on and off—use the fader to gradually bring it in, then gradually take it out. Easy to do and sounds much more professional.

Lastly, the final BOOOMMM. Well, you know what happened there, right?

When you just can't get enough

There are lots of great resources that can provide different perspectives on what we've discussed while explaining things in greater detail. I'll list a few here that I can endorse wholeheartedly (hoping they don't disappear by the time you read this…such is the nature of the industry).

The main thing you can do to improve your knowledge and skills at audio is to find a good mentor and practice. Practice. Practice. It simply takes time and practice to get good at something, so start now and get on with it. Meet with other audio folks at local churches once in a while to share ideas. Partner with other churches to hire a professional trainer to come work with you for a weekend. Attend a good worship conference where you can take classes and, even better, spend time getting to know other individuals who run into the same issues you face each week. If this is your contribution to God's work, then it's your responsibility to develop your skills as much as you can. The best artisans and skilled workers were hired to build the Temple, and I believe He deserves the very best we can offer.

Resources to check out:

www.worshipMD.com — Doug Gould seems to know everyone in this business, and as a former high-level manager at Shure, Tascam, and E-Mu Systems his mission in life is to enable everyday folks like you to serve your church's audio needs. He's a consultant who travels the country organizing worship conferences and teaching classes.

www.churchtecharts.com — Mike Sessler is a long-time industry veteran who has lots of helpful advice on mixing, gear, and people issues in church. His excellent podcast, *churchtechweekly*, brings together several industry experts that share experiences and insights into a wide range of topics.

www.goingto11.com — Dave Stagl is another highly-regarded expert who has served churches as well as operated as an independent consultant, offering training sessions for churches, teaching classes at conferences, and mixing special events.

The Ultimate Church Sound Operator's Handbook, Bill Gibson — You want more detail? This book has it in droves and is an excellent companion to the book you're holding right now. It also features a DVD with lots of examples to illustrate concepts in the text.

https://www.facebook.com/groups/worshipaudiocollective/ — There are lots of Facebook groups out there for audio, but this one features real professionals who won't steer you wrong…don't trust everything you read on the internet…

About the author

Dr. Barry R. Hill is professor of music and director of the Audio & Music Production degree program at Lebanon Valley College in Pennsylvania. A member of the National Academy of Recording Arts and Sciences (Grammys), Audio Engineering Society, and the Themed Entertainment Association, he has extensive experience in the industry as a recording engineer, consultant, conference panelist, workshop presenter, and performer. As Director of Worship Arts for his church, he has spent years designing and installing sound systems, consulting on church acoustics, and training sound teams. His doctorate is in learning & instructional design, which he uses to help teachers design better experiences for students. For fun he teaches theme park design concepts in the college's Experience Design program, applying a life-long interest in Disney Imagineering and theme parks.

Dr. Hill holds degrees in Instructional Design from The Pennsylvania State University, Music Technology & Interactive Media from New York University, and Music with Recording Arts from the University of North Carolina Asheville.

www.ingramcontent.com/pod-product-compliance
Lightning Source LLC
Chambersburg PA
CBHW070236190526
45169CB00001B/197